WRITTEN IN BLOOD

www.transworldbooks.co.uk

WRITTEN IN BLOOD

Mike Silverman with Tony Thompson

BANTAM PRESS

LONDON · TORONTO · SYDNEY · AUCKLAND · JOHANNESBURG

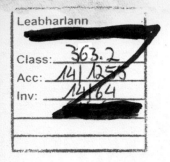
TRANSWORLD PUBLISHERS
61–63 Uxbridge Road, London W5 5SA
A Random House Group Company
www.transworldbooks.co.uk

First published in Great Britain
in 2014 by Bantam Press
an imprint of Transworld Publishers

Addresses for Random House Group Ltd companies outside the UK
can be found at: www.randomhouse.co.uk
The Random House Group Ltd Reg. No. 954009

The Random House Group Limited supports the Forest Stewardship Council® (FSC®),
the leading international forest-certification organisation. Our books carrying the FSC
label are printed on FSC®-certified paper. FSC is the only forest-certification scheme
supported by the leading environmental organisations, including Greenpeace. Our paper
procurement policy can be found at www.randomhouse.co.uk/environment

Typeset in 12½/15pt Bembo by
Kestrel Data, Exeter, Devon
Printed and bound in Great Britain by
Clays Ltd, Bungay, Suffolk.

2 4 6 8 10 9 7 5 3 1

To Sam, for all the time she gave me to think

1

---

27 January 1984

HE WAS LYING ON HIS SIDE, ONE ARM DRAPED CASUALLY ACROSS his stomach, the elbow of the other sticking out at an angle, as if he were about to use the hand to prop up his chin. Had it not been for the vast gash that had, to all intents and purposes, separated his head from the rest of his body, and the copious amounts of blood that covered the walls, the floor and almost every inch of his clothing, you might have thought he was simply taking a nap.

The body was in the narrow hallway of a third-floor flat in Islington, north London, one of those red-brick blocks where the front door opens out on to a shared balcony-cum-landing. As an expert in blood pattern analysis with the Metropolitan Police Forensic Science Laboratory (MPFSL), it was my job to provide an insight into the circumstances surrounding the man's death, ideally as quickly and efficiently as possible.

The victim's brother had stumbled across the shocking scene a little earlier that morning and had immediately called the police. Uniformed officers arrived within minutes and sealed off the flat before informing the CID, who, after a quick visual inspection from the front door, requested the services of a blood pattern expert and sent a car to collect me from the lab's Lambeth base.

Whenever a forensic scientist is called in on a murder case, they never know quite what they are going to encounter, or how easy it's going to be to come up with anything useful. Time is always of the essence and, as no one else can carry out further examination of the scene until they've finished their part, the pressure to produce results is always intense.

As I stepped forward, the detective superintendent, police photographer, scenes of crime officer (SOCO) and my assistant crowded around the open doorway so that they could see into the hallway and watch me work. I could feel the weight of their anticipation on my shoulders.

I moved slowly, wary of standing on any potential evidence, my plastic overshoes sticking slightly to the still-tacky blood on the floor, my eyes scanning the scene for familiar shapes and patterns in the blood – the tell-tale marks from the victim or their killer which would enable me to reconstruct the sequence of events. (Usually, after initially accessing the scene, one of the first things a forensic scientist would do is to carry out tests on visible stains and fluids to get a better indication of whether they are indeed dealing with blood, but in this case that seemed unnecessary.)

The victim appeared to be of Chinese or Vietnamese origin, and I didn't need any special training to surmise that the weapon had most likely been some kind of cleaver – the preferred weapon of the underworld society known as the Triads.

I moved forward, carefully avoiding the many clumps of hair that had fallen to the floor, having been cut away as the blade first sliced into the man's neck. I squatted down in the narrow space alongside the body to take some pictures with the Polaroid camera I was carrying. The right hand had been severed at the wrist and several fingers were missing from the left. The man's eyes were open, staring blankly into the distance.

The wound in his neck stretched all the way across and was as deep as it was wide, causing his head to loll forward at what would otherwise be an impossible angle. The ragged flesh was speckled with shiny black dots of congealed blood, contrasting with the sharp whiteness of his exposed vertebrae. As the victim had been dead only a few hours, there was little decomposition and, thankfully, no smell, other than the heavy, metallic scent of the blood itself.

You never get used to dead bodies – especially in those cases where the victims are children or babies, or when the body is infested with maggots or the insides have turned to soup – but you have to be professional, you have to put it out of your mind and simply do your job. It sounds callous, but you can't allow yourself to think about the body as a person; you just have to see it as the source of the blood that is all around you, the point of origin of the distribution.

Although, to the untrained eye, one splash looks very much like another, individual blood patterns can be highly distinctive and can be produced only by specific events. Blood is a remarkable substance: no other liquid on earth behaves in quite the same way. Analysing the patterns it leaves behind at crime scenes requires a detailed knowledge of fluid dynamics, mathematics and biology, and, more often than not, the way blood behaves is counter-intuitive.

Once you learn to recognize a particular pattern, it's relatively easy to work out what must have occurred in order for it to have been created. The size of individual bloodstains can tell you how much energy was involved in their production, but while you might expect smaller, lighter blood droplets to travel the greatest distance from the point of origin, it's actually the larger, heavier drops that travel furthest.

Moving droplets of blood form tapered stains which 'point' in the opposite direction to the one you would expect. Bloodstains also vary enormously depending on the surface they strike. A drop of blood hitting a polished, smooth surface from a height of 80 feet will form a perfect circle on impact. The diameter of the resultant stain is determined not only by the height from which the drop has fallen but also by the geometry of the object from which the blood drips. For example, blood dropping from the point of a knife will result in small volume drops, whereas blood dropping from the edge of a saucepan will result in much larger volume drops.

The same-size drop falling from just 18 inches on to a slightly dusty but otherwise smooth surface will form a radial satellite pattern that spatters in all directions. Then there are the patterns caused by blood dripping into more blood, blood that is cast off from a moving weapon, spattered by the impact of a weapon into wet blood, or blood coughed up from the lungs. In certain circumstances, red blood cells in pools of blood can gather together, leaving a clear serum around the outside edge and making it look as though someone has tried to clean up the stain, even though no one has touched it.

Blood isn't just a liquid, of course, it's a complex mixture of plasma, cells, enzymes, proteins, antibodies, antigens and other components. Blood can congeal, agglutinate or clot, all of which have different causes. Blood congeals as a result of an environmental change such as cooling; agglutination, where cells in the blood clump together, is caused by the reaction of antibodies and antigens; clotting is the coagulation of blood as a result of the action of blood clotting agents in order to heal wounds.

During the course of my training, my colleagues and I re-created and studied hundreds of patterns and soon realized that, when it came to experiments, only blood would do. We had tried using red ink and other coloured liquids, but the surface tension and viscosity was never right, so the results were never quite accurate. We sometimes made do with pig or horse blood for our tests, but our best results were always obtained when we used (out-of-date) human blood obtained directly from the transfusion service, with which we had a special arrangement. Another advantage of using this blood was that it was free from disease, thanks to the extensive screening process the transfusion service has in place.

Having worked my way through pints of the stuff in order to reproduce dozens of criminal scenarios, visiting a crime scene was then a matter of matching what I saw with what I knew.

Today, the hallway floor was covered with the victim's blood, but that didn't tell me much. Volume rarely does. I'd had cases where only a few splashes of blood were present and these had given me the full story, while, in others, pints of blood had told me nothing at all.

At one particular job in Ilford I'd been called in to examine a domestic kitchen that was virtually awash with blood, though no bodies had been found. There were no specific marks or patterns anywhere, but the sheer volume of blood was so great that my initial impression was that the police were looking for at least two murder victims and that their bodies would have been totally drained.

It turned out that not only was there only one victim, but he was still alive and, even more remarkably, despite such blood loss – from, it turned out, a deep head wound – he had managed to drive himself to the local hospital.

By contrast, there were several distinctive features in the blood patterning in this case, so I knew at once that I had plenty to work with. On the wall just above the man's left shoulder I could see a series of dark-red splashes that looked a little like upside-down musical notes or tadpoles – fat heads with long, skinny tails that trailed down towards the skirting board. This distinctive pattern is created by blood spurting out of an artery that has been severed while the victim is still alive. In such a situation, the blood emerges under pressure, in large, heavy globules. Here, each series of fat heads on the wall represented a beat of the victim's heart; and the tails had been created by gravity pulling the excess blood towards the ground. The stains formed a number of downward arcs, displaying the double beat of the heart and indicating that each new spurt had been slightly weaker than the one preceding it, as the victim's blood pressure dropped and his heart slowed.

Further up the wall, at about shoulder height, I could see another characteristic pattern, this time caused by the impact of something solid into wet blood. This mark must have been made prior to the blow that severed the victim's artery. Above this, high on the walls and across the ceiling, I could see several long, thin trails of blood. These I immediately recognized as weapon tracks.

If you've ever cut your finger while using a sharp knife in the kitchen, you'll know that, for the first fraction of a second, you don't see any blood. This is because the immediate reaction of the body is to close the capillaries (the very small blood vessels that take the oxygenated blood to the cells of the tissues of the body). The same principle applies to most wounds, regardless of the weapon used. The first time you strike someone hard enough to break their skin, for a split second, no blood flows. It's quite possible to plunge a knife deep into a person's body, or slash with a razor, and have the blade emerge without any visible traces of blood. It's usually not until the second blow that the victim's blood is flowing and is deposited on the weapon. As the attacker pulls back for each subsequent strike, blood that is on the weapon is flicked off, leaving a linear track mark.

Yet more patterns on the walls were formed by impact spatter, caused by the blade of the cleaver striking wounds already soaked with blood which had then splashed off in various directions.

By the time I reached the far end of the hallway, I'd seen enough to be able to draw my initial conclusions, and called out my findings so that my assistant could take notes and the detective superintendent would know what was going on.

'The victim was standing up when the attack started,' I began. 'It looks like there were at least two blows in that position, with the victim just inside the hallway before he collapsed to the ground.' I was speaking slowly, choosing my words every bit as carefully as I was choosing my steps. At such an early stage, a rushed interpretation or vague explanation can send the investigation spiralling off in the wrong direction.

'He was already lying down, close to the position he's in now, when the neck was severed,' I continued. 'That's where the majority of the assault took place.'

I asked the superintendent to close the door to the flat so I could take photographs of the other side of it and see if any more blood was present. This might tell me whether the door had been open or closed during the attack and help me work out where the killer had been standing in relation to the victim.

A second after the door was closed, the flap of the letterbox was lifted and the eyes of the superintendent appeared.

'One thing. As you are the first person to go inside, I thought I should mention that the flat hasn't been searched yet. It's possible that the killer is still inside. Just wanted to let you know.' The flap snapped shut. I was squatting down on my haunches – almost sitting – directly on top of the victim's chest, trying to get the best angle for my photographs, and immediately I felt myself tensing up.

I looked behind me. The hallway was L-shaped, and there was a thin smear of blood present on the far back wall, showing that someone had been there at some point. With no other blood trail on that part of the floor, the smear was more likely to have been left by the attacker than the victim.

Part of me wanted to leave the flat there and then and have a police officer come in and make sure no one else was there, but that would have risked compromising the crime scene. There was simply no way to tell if someone was hiding here or not. For all I knew, there could even be another body back there.

At that moment, I looked down at the victim's legs, which were crossed at the calves, his feet clad in grey socks, and saw that blood streaked along the floor between them and the door. These were drag marks, indicating that the body had been pulled back from the doorway by a foot or two in order to allow the front door to open. That in turn meant that the killer had, more likely than not, already made their escape.

I breathed a sigh of relief and carried on with my work. I planned to take a few more photographs and then swap my camera for a clipboard so I could start drawing sketches.

It might seem odd that, even today, crime scenes are still drawn by hand, but the simple truth is that there is a limit to how much information can be obtained from a photograph. It is not possible to tell which stains are in fact blood or indeed to see the smaller, most significant bloodstains or capture the relationships between the various items seen from different perspectives.

That's why, once the initial photographs are taken, a scene examiner also needs to take measurements and draw sketches which will allow them to include all significant bloodstains and patterns but exclude staining and objects that are mean-ingless in a way a photograph cannot. (See pages 2–4 of the first picture section to compare my original photographs and sketches of the scene.)

The other thing I was looking out for during my examina-tion was the presence of so-called 'alien' blood – blood that might not belong to the victim.

In many violent crimes, and in stabbings in particular, the assailant often ends up injured and leaves their own blood at the scene. This is especially common where the weapon used is a kitchen knife. Unlike hunting knives, domestic knives do not have a quillon, or finger guard, between the blade and the handle, as they are not designed for stabbing. So if such a knife is used in a stabbing motion and hits bone, the attacker's hand is likely to slip forward on to the blade itself, cutting the holder's fingers or palm.

Such injuries rarely happen when cleavers are used, and these weapons also allow for multiple slashing strikes all across the body in rapid succession, which has a symbolic significance – members of Triad gangs take an oath of loyalty in which they agree to die 'by a thousand cuts' should they betray the brotherhood.

In TV dramas featuring the work of forensic scientists, there is almost always a eureka moment in each episode where the boffin finds the one piece of evidence that solves the case. In reality, such moments are few and far between, but today, as I carefully manoeuvred my way around the nearly headless body in this narrow hallway, I was about to experience one.

Lying on the floor by the door was a small peaked cap, the kind that would be worn by a schoolboy. It was tucked into the corner behind the door and had been hidden from view while it was open, which is why I hadn't noticed it before. I made my way towards it, taking photographs of it *in situ* before examining it more closely.

I looked back over at the victim. The bloodstain patterns on both sides of the hallway made it clear that he had not moved much after the first strike. Aside from rolling, or being rolled, one way and then the other – both sides of his body were heavily bloodstained – he had gone down close to where he now lay.

It was possible that he'd been attacked soon after answering the door, though at some point the killer had been behind him, at the far end of the hall. The position of the cap, pushed back behind the door, suggested that it may have belonged to someone other than the victim, and that made it a highly significant find.

'There's a cap here,' I called out, raising my voice to be heard through the closed door. The flap of the letterbox immediately sprang up. 'Is there?' asked the superintendent. 'I don't suppose it's got the name of the murderer in it?'

I turned the cap over to look at the inside and saw a strip of white tape that had been sewn on to the lining and bore a handwritten name in neat, slightly faded but still perfectly legible block capitals. 'Funny you should ask that . . . there *is* a name.' I read it out to him.

The superintendent asked to see the cap, and it turned out to be the key to solving the case. I may have been the one who found the cap, but in this case there was precious little forensic science involved in finding the attacker.

The name inside the cap was that of the schoolboy to whom it had belonged. The investigating team tracked down his family, and it turned out that the cap had been sold at a church jumble sale a few weeks earlier. Remarkably, the boy's mother not only remembered the powerfully built oriental man who had bought it but was able to pick him out in a line-up.

It was an incredibly lucky break and virtually the only clue. Back then, it would still be several years before DNA became a tool forensic science could make use of, so, without the name in the cap, there would have been little to go on.

Even with a suspect in custody, however, the work I was doing at the crime scene was crucial. If, for example, the suspect claimed that he had struck out in self-defence, that someone else carried out the attack and that he was only a witness or had tried to intervene, the blood pattern evidence could indicate either that he was lying or support his version of events.

Interpretation of bloodstains makes it possible to reconstruct a sequence of events with far more accuracy than even the memory of an eyewitness can give, and in cases where there are no witnesses, the splashes, smears, swipes and wipes that are left behind allow the deceased to show you exactly what took place during their final moments. It's as if the details are written out on the walls in letters of blood.

In all, I spent four and a half hours at the crime scene, negotiating my way gingerly around the victim, before I felt I had enough material to support my conclusions, had completed my drawings and marked up all the stains I wanted the scenes of crime officer to take samples of for me.

I could have taken the samples myself but to do so would require me to write a 'chain of custody' statement to prove their continuity of possession between the scene of the crime and the lab. This might subsequently require me to appear in court merely to confirm that continuity – something that would keep me out of the laboratory for the whole day and disrupt my work on other cases. By getting the SOCO to collect the samples instead, I could ensure I would need to make only a single court appearance, to confirm and explain the results of my scientific findings.

Once all the desired samples had been properly photo-graphed, each with a scale to show the size of the mark (and possibly an arrow to indicate its orientation), these selected stains would be scraped or swabbed off the walls, or, if they were on wallpaper, cut out and collected in sealed polythene bags. The samples would then be taken back to the laboratory, where further testing would take place.

By this time, the forensic pathologist had arrived to examine the scene before requesting the transport of the body to the local mortuary, where a forensic autopsy would be conducted. Normally, I would attend the autopsy personally, as learning about the cause of death can help with the interpretation of bloodstains, but, as was the case with testing for blood in the first place, this seemed utterly unnecessary, as the cause of death was obvious.

I made my way out of the flat and headed towards the nearest bus stop – the urgency had gone, so police transport was no longer available – to make my way back to the lab. In those days – early 1984 – the Metropolitan Police Forensic Science Laboratory was the most sophisticated facility of its kind in the world, and the equipment I was using and techniques I had at my disposal were all state of the art.

Having started at the laboratory some six years earlier, I had finally reached the point where I felt that I had mastered all the skills I needed to be fully able to do my job. However, all that was about to change. Within just a few months there would be a major scientific breakthrough that would confirm Britain's position as a world leader in the field of forensic science. It would change virtually every aspect of my work and would also, ultimately, usher in the beginning of the end for the Forensic Science Service, leading the organization to shut its doors for the very last time in March 2012.

All that lay ahead, but as I waited at the bus stop and a chill January rain began to fall, I couldn't help thinking back to the start of my career and pondering the fact that, had it been raining back then, I'd probably never have become a forensic scientist at all.

2

_____

AS FAR BACK AS I CAN REMEMBER, I NEVER HAD ANY INTEREST IN becoming a forensic scientist.

I grew up in the east London borough of Hackney during the fifties and, though mine wasn't a particularly academic household – my dad was a tailor, my mother a secretary, and neither had been to university – I did well enough at the 11-plus to apply for and subsequently win a full scholarship to a prestigious private school in the City of London.

Not surprisingly, my parents had plenty of ambition for me. They knew that attending such a school would significantly boost my chances of getting a degree from the likes of Oxford or Cambridge and hoped that this would set me on the path to a life of success. As a result, we were all hugely disappointed when my scholarship was withdrawn when it was discovered that my parents had unwittingly breached its terms and conditions by moving to Ilford – well outside the scholarship catchment area – just a few months before I was due to enrol.

I ended up at the local grammar school, where at least my parents could afford the uniform, and where I enjoyed science and discovered that I had a minor talent for drawing – which came in useful when sketching the results of my biology dissections. Languages were a major weakness and I was required to leave my O-level Latin class after mistranslating '*Quo vadis?*' ('Whither goest thou?') during a test – the equivalent of spelling your own name incorrectly on the exam paper. I thought 'Oi, where have you been?' was close enough.

Although not as highly regarded as the private school I had hoped to go to, the grammar school still expected the majority of its students to end up at university or polytechnic, and a significant number were planning to become either lawyers or doctors. I had no particular career in mind but, as I had to choose something, I opted to join the ranks of the latter.

I got an interview for the medical school at what was then known as the London Hospital, in Whitechapel, but it didn't go particularly well. Also, at the time, Latin was an unofficial prerequisite for anyone wanting a career in medicine, so when the personnel department found out I didn't have it, they upped my required grades to far higher than I was likely to achieve – a way of rejecting me on the sly. (Ironically, many years later, the same hospital – now renamed the Royal London Hospital – would offer to name a pathology-teaching wing after me.)

With a medical degree out of the question, I opted to take a course in either chemistry or applied biology and received offers from Queen Mary College and Northeast London and Thames polytechnics. Unable to choose between them, I decided to leave it to fate. If the weather was sunny on enrolment day, I'd hop on the Woolwich Ferry and make my way to Thames Poly. If, however, it was raining I'd take the Tube and sign up for Queen Mary. When the day came, it was gloriously sunny, so I took the ferry.

In common with most polytechnic courses at the time, the third year of my sandwich course was to be spent working in industry. Some students had managed to set up industrial placements with the Medical Research Council or the Royal Botanical Gardens at Kew – anywhere in need of student biologists to carry out some research – but quite a few of us were struggling to find a suitable position.

As a last resort, tutors stepped in to help by calling in favours from friends and former colleagues. The dean of the biology faculty at Thames had been at college with the man who was then head of the microbiology facility at the MPFSL – which is how I ended up spending a year working as a research student at the laboratory.

In September 1974 myself and a classmate named Phil turned up for our first day and quickly discovered that we were only the second and third students the laboratory had ever taken on, so they weren't entirely sure what to do with us.

After being kitted out in white laboratory coats and intro-duced to the other members of the serology department (the department responsible for analysing blood and other body fluids) where we would be working, we embarked on several weeks of training in the many serological techniques used at that time in forensic blood grouping. After a while, I was given my research assignments: to develop a new plant 'antiserum' (a phytohaemaglutanin or lectin), from grape seeds and another from the albumen glands of the common garden snail.

The science of serology is a relatively recent one. Although blood is by far the most common body fluid left at crime scenes, it was, until the twentieth century, impossible even to distinguish human blood from animal blood, let alone the blood of one human being from another. This led to a large number of killers escaping justice, as they could claim that blood found on their hands or clothes got there while they were skinning a rabbit, or was the result of a nosebleed, for example, and no one could do anything to prove them wrong.

If the blood had been allowed to dry, it couldn't even be accurately identified as blood. Although red when wet, dried blood can turn a brown colour and, to the untrained eye, looks similar to a number of other substances, such as paint, chocolate and woodstain. Using a microscope, it was possible to identify liquid blood because of the distinctive doughnut-like shape of the red blood cells it contains, but these cells break down and their shape disappears when the blood dries and do not return even if the stain is rehydrated.

In the 1800s, finding a test that could positively and reliably identify blood 100 per cent of the time was considered the Holy Grail of criminal medicine. Various experimental procedures were attempted, but none proved satisfactory, as they generated too many false positives.

One man who was fully aware of the significance of such a test was Sir Arthur Conan Doyle, creator of the celebrated cerebral detective Sherlock Holmes. In fact, the very first time Holmes makes an appearance in print – a few pages into the first chapter of *A Study in Scarlet* (published in 1887), his first words have him finding a solution to this very problem: 'I've found it! I've found it. I have found a reagent which is precipitated by haemoglobin, and by nothing else.' When Dr Watson questions the practical application of such a discovery, Holmes is more than happy to enlighten him:

Had this test been invented there are hundreds of men now walking the earth who would long ago have paid the penalty of their crimes . . . Criminal cases are continually hinging upon that one point. A man is suspected of a crime months perhaps after it has been committed. His linen or clothes are examined and brownish stains discovered upon them. Are they blood stains, or mud stains, or rust stains or fruit stains or what are they? That is a question which has puzzled many an expert, and why? Because there was no reliable test. Now we have the Sherlock Holmes test, and there will no longer be any difficulty.

Curiously, the test is never mentioned again anywhere within the Sherlock Holmes canon and it would be more than a decade before such a procedure became a scientific reality.

Until 1628, when English physician William Harvey discovered that blood was circulated around the body via veins and arteries by the pumping action of the heart, a leading school of thought was that blood was manufactured in the liver from food in the stomach and then transported to the heart, where it was 'burned' to provide energy.

For this reason, the earliest – always unsuccessful – attempts at transfusion usually involved patients being asked to drink blood. Once the functions of the veins and arteries were understood, transfusions using first animal then human blood were made intravenously, but this, too, proved problematic. Several patients survived procedures involving small amounts of sheep or calf blood in the late 1600s but, once larger volumes were attempted, the death rate soared, which led to transfusions being banned in Europe for more than 150 years.

Experiments did resume, but the risks remained high. In 1871, for example, 263 human-to-human blood transfusions were carried out, and 146 of the patients died. This survival rate of roughly one in two led to the not unreasonable supposition that human blood was not all of one type: there were at least two different kinds and these were biologically incompatible. However, it would not be until 1901 that scientists fully understood the make-up and functioning of the blood, thanks to research carried out by Karl Landsteiner, who at the time was an assistant professor at Vienna University.

Red blood cells, one of the key components of blood, contain surface molecules called antigens which are responsible for controlling immune reactions through the production of antibodies. In turn, antibodies identify and assist the destruction of bacteria, viruses and other foreign cells in the bloodstream. Antibodies will also attack red blood cells, such as those introduced during a transfusion, if they have antigens different to those of the host, and this reaction often proves fatal.

Experimenting with his own blood and that of five of his colleagues, Landsteiner observed that certain mixtures resulted in agglutination (to recap, the clumping together of the red blood cells). He drew up a cross-referenced table that suggested the existence of three separate types of blood: A, B and C (subsequently renamed O.)

It soon became clear that transfusing a subject with type A blood (and therefore A antigens on the surface of their cells) with blood from someone who was type B (but with anti-A antibodies in their serum) would cause agglutination within the subject's body and would kill them if they were given a large enough transfusion. However, someone with type O blood had neither A nor B antigens on the surface of their red blood cells, so their blood could be transfused into anyone without causing any harm.

The following year, a fourth blood group, AB, was added to the list and a full blood compatibility chart drawn up. People with type O blood are known as universal donors, as their blood can be given to anyone in transfusion. However, they can only receive blood from other type O individuals. Those with type AB blood have both A and B antigens on the surface of their red blood cells. They can receive blood from members of any other group so are also known as universal recipients, but they cannot donate to those with type A, B or O blood, due to the presence of incompatible antigens.

Around the same time that Landsteiner was carrying out his research, German scientist Paul Uhlenhuth was conducting a series of experiments which involved injecting a small amount of human blood into a rabbit, or other mammal, and waiting for the animal's immune response system to produce the specific antibodies needed to fight the perceived threat.

Uhlenhuth would then draw blood from the animal and separate out the cells to create a serum that was rich in anti-human antibodies. If this serum was mixed with a sample of human blood, a visible white solid (a precipitate) would be produced as a result of the antigen/antibody reaction. If the test sample did not contain human blood, no reaction would be seen. Helpfully, the test worked just as well on dry blood as on the liquid form.

By injecting rabbits with blood from dogs, cats, deer, cows, sheep and other creatures, Uhlenhuth was able to produce a series of sera that could be used specifically to identify exactly what type of blood an investigator was looking at. After a long wait, science suddenly had access to tests that could not only distinguish human blood from other substances but could tell the blood of one subject from that of another. This was the beginning of modern forensic science.

There were, however, some limitations. Uhlenhuth's test could not distinguish between genetically similar animals, so, for example, the blood of a human would produce the same reaction as the blood of an ape. However, as crimes involving apes were few and far between, this restriction did not cause much concern. A bigger problem was the fact that, with just four blood types, each type could be found in large proportions of the population.

Population studies carried out over the decade following the discovery of blood types showed that, within the UK, around 42 per cent of people were type A, 10 per cent type B, 45 per cent type O and just 3 per cent type AB. Tests further afield showed that blood type distribution varied significantly from one country to the next. In Taiwan, for example, around 40 per cent of the population are type O, while among the Ainu people of the Japanese island of Hokkaido just 23 per cent have type O blood.

Over the years that followed Landsteiner's discovery, the four blood groups were split into further subdivisions as a result of protein analysis that separated out distinguishing factors or combinations of factors among smaller categories of the population.

In 1925 Landsteiner realized that all body fluids – notably, sweat, saliva, urine and semen – could be similarly typed in the case of around 86 per cent of the population. These people became known as secretors, as their body fluids also contained blood group antigens. The 14 per cent that did not have such antigens in their other fluids were designated non-secretors.

It was also discovered that blood groups are inherited. This proved useful in early paternity testing, as certain parental combinations are excluded from the progeny of a particular blood group. For example, a blood type AB father cannot have a type O child.

By the late seventies blood grouping had moved on enormously from the time of Landsteiner, with upwards of a dozen different blood grouping systems being in routine use at the MPFSL. ABO blood grouping was still the principal way of distinguishing the blood of one person from that of another but, with such a high proportion of the population sharing the same blood group, more refinement was needed to make the system more useful.

New research was constantly being conducted to find additional sources of antibody-like substances that could be used to type bloodstains or determine subgroups of blood types. My research project involved attempting to develop certain plant extracts from grape seeds (and, later, albumen glands from the common garden snail) which would react with certain specific blood types in stains.

Before we could start work, the laboratory staff needed to check that we could perform basic procedures, so they asked me and Phil to repeat some of Landsteiner's early experiments by taking blood containing A cells and mixing it with the correct antigens in order to get the sample to agglutinate. Almost immediately, Phil ran into problems and, after this, whenever he carried out this procedure, it just wouldn't work. No matter what he tried, he could never get blood to agglutinate. It was almost as if there was something in him that made the experiment fail. At first, some of the staff thought he might be doing something deliberately wrong, but he wasn't, and no one could explain why it was happening. We tried mirroring each other's tests. Mine would do exactly what it was supposed to do, and his would not work at all. Eventually, Phil had to leave and continue his year in industry elsewhere. He hadn't done anything wrong; it was simply that if he couldn't make such a basic reaction work, for whatever reason, there was just no way anyone would be able to trust whatever results he came up with during the course of his subsequent research. I was left to carry on the research alone.

I soon found out that there was a degree of internal rivalry between sections within the biology division. The staff of the blood grouping sections where I worked considered themselves to be pure scientists, while those in the 'general search' section, who went out to crime scenes, examined exhibits, wrote up statements and went to court were considered little more than administrators. And if you spoke to the reporting officers – the scientists who went to court to present their findings – they would tell you that they were the ones who did all the clever forensic stuff and the people in grouping who did the actual scientific analysis were merely technicians.

My work involved looking into substances known as lectins and phytohaemoglutanins that could make specific human blood cells agglutinate but which came from plants rather than animals. It was known that there were sub-categories of ABO types in blood, but these could not be determined in dried stains. I was to look for such haemoglutanins that would work with dried bloodstains.

Initially, I worked with grape seeds. A scientific paper written in Japan had reported that a type of phytohaemoglutanin was present in the thin brown coats of the seeds and that it had the ability to agglutinate blood cells of all four types. However, by growing a certain mould on an extract of the seed coats, you could create phytohaemoglutanins that agglutinated only type B blood cells. If I could successfully reproduce the results from Japan, it would create a ready source of cheap and specific anti-B serum.

My experiments worked, and the resultant material did indeed react with and agglutinate human type B cells. But the phytohaemoglutanins took weeks to produce and the procedure was incredibly fiddly and generated only small amounts of usable material from sackloads of grapes. I moved on.

My second source of phytohaemoglutanins was far more promising – those albumen glands of the common garden snail – though the process of extracting it was anything but pleasant. I found bucketfuls of likely specimens behind the garden shed of my then girlfriend's house in Bishop's Stortford, gathered them up and set about preparing them.

The first step was to kill the snails, which I did by placing them in a large jar and putting a cloth soaked in chloroform on the top. Although this killed them humanely and effectively, it also caused them to exude a thick green defensive slime, which I had to pick them out of before I could move on to the next stage. Each snail had to be dissected under a magnifying lens and have its albumen gland removed. When I had enough glands, I placed them into a blender and mixed them together, before moving on to extract the phytohaemoglutanins. It was, in all, an extremely messy process.

When I applied the final extract to bloodstains, I discovered that not all type A bloodstains were identifiable through this method. It became apparent that the extract worked only with a subgroup known as type A1 cells and so could be used to distinguish between A1 and A2 type cells in bloodstains. This was exciting stuff. There wasn't at the time – and possibly still isn't – any method of distinguishing subtypes of group A in stains, so this development was novel and useful, as it would improve the discriminating power of ABO bloodstain grouping.

The more that the percentages of population could be broken down according to blood type, the more effective the judicial system could be. As a student conducting my first ever piece of research, my results were more than I could ever have wished for and, for a short time, I was my supervisor's blue-eyed boy. It seemed inevitable that my research would soon be published in a leading scientific journal.

It would, of course, be necessary to demonstrate the robustness and reproducibility of my results before it would be possible to publish or introduce the new technique into the armoury of blood grouping methods available to the forensic scientists in the lab. All I had to do was mix up another batch of the extract and the trials could begin.

I made the second batch using an identical method to the first, right down to the species of snail, but when I tried out the

new extract on control bloodstains, the expected differentiat-
ing reaction failed to occur.

After trying to reproduce the initial results for several months
without success, using snails from here, there and everywhere,
someone suggested that, rather than being anything to do with
the albumen gland itself, the desired anti-A1 reaction could
have been related to the type of plants the original snails had
been eating, as it had been shown that these substances collected
in the gland.

With this in mind, I went back to my girlfriend's house to
gather up more of the original snails, only to find that the
garage had been taken down, the ground concreted over and
that were no snails to be seen anywhere. It was then I realized
that I was never going to be able to tell why it was that the
technique had worked. My original extract continued to be
effective but could not be reproduced, and within a few weeks
I was back at college to finish my degree. There was to be no
publication or Nobel Prize.

I spent the summer after my graduation driving a van and
delivering kitchen furniture while waiting to see whether
my degree was good enough for me to be able to find a
postgraduate research post. By the time I got my results, all
the most interesting research posts had already been taken and
the only funded PhD on offer had the compelling title 'An
Investigation into the Alternative Respiratory Cycles of Red
Bread Mould'. I decided to accept.

Soon after I began, my research supervisor left on the first
of two virtually back-to-back periods of extended maternity
leave. Consequently, I had little contact with her, no supervision
and few colleagues interested in fascinating dialogues on the
wonders of bread mould of any colour. I became increasingly
disillusioned and found myself questioning the value of what I

was doing when the local mayor came on a tour of the research laboratory where I was based. She asked what I was working on and when I told her, she said, 'That's going to be extremely useful for the baking industry.' It wasn't, of course; it was just that the only word she had picked up on was 'bread' and she assumed that I was looking for a way to stop bread spoiling so quickly.

I'd kept in touch with people from the Metropolitan Police Forensic Science Laboratory since my year in industry. In particular, I would meet up with head of research Pete Martin and others from the team almost every Sunday for a lunchtime drink. I knew that very occasionally vacancies came up in the laboratory but, having only experienced the blood grouping section, I didn't at the time have a real understanding of what forensic science was all about and wasn't looking for a career in that field.

Then, one day, almost out of the blue, Pete Martin called and invited me to apply for a job as a scientific officer at the MPFSL. I was sufficiently disenchanted with the progress of my postgraduate research to wish to take him up on the offer but I also felt that it was important to finish my PhD first.

'The thing is, the job's available right now and, if you wait, it will go to someone else,' Pete told me. 'Anyway, we don't take on people as scientific officers if they have PhDs. They have to come in at a higher pay grade and we're not recruiting to that grade any more. Tell you what, if you get the job, give it a few months and then we can give you day release to finish your studies.'

I was more than happy to accept the compromise.

'Just one last thing,' said Pete. 'It's not in blood grouping, it's in general search.'

My first instinct was to turn it down. Although I was very

familiar with the work of the grouping section and had enjoyed my time there, I knew nothing about general search and didn't really know anyone who worked there. In the end, it was Pete Martin who persuaded me to give it a try.

'See how it goes. You can always move around later on,' he explained.

For my first few months, I let my research studies go completely so I could focus on getting to know my new job. After that, I tried for a little while to balance work and study, but it soon proved impossible. I spent most of my time at my weekly day release at the research laboratory recovering the equipment that had been 'borrowed' by undergraduates during the previous week. I scarcely had time to conduct any experiments. It didn't take long, however, for me to realize that forensic science was exactly the sort of work I wanted to do and that the chance to do it full-time was far more important than the chance to have some additional letters after my name.

The lack of a doctorate would come back to haunt me on a few occasions in the future but, there and then, I was in absolutely no doubt that I had made the best possible decision with regard to my future.

3

THE METROPOLITAN POLICE FORENSIC SCIENCE LABORATORY
that I was then about to join as a full-time scientist may have
been a state-of-the-art facility and the envy of police forces
around the world, but the service it provided hadn't always
been so highly revered. In fact, had it not been for the out-of-
hours hobby of a certain Cyril Cuthbert, a humble sergeant in
charge of the pay ledger at a south London police station, the
laboratory might never have existed at all.

Originally from Devon, Cuthbert had initially hoped to
work in the medical field but gave up on both medical and
dental training and joined the Metropolitan Police in May
1924 in order to pursue a career in law enforcement. Having
joined the police, however, his interest in science continued
and he began taking evening classes in chemistry and reading
books by leading pathologists of the day such as Sydney Smith
and renowned chemist Ainsworth Mitchell. When colleagues
in the force learned of his interests, they began to consult with
him about cases that were causing them difficulty and he was
more than happy to help out as best he could.

Increasingly intrigued by the various ways in which science could assist in solving certain crimes, Cuthbert saved up a month's salary in order to purchase a second-hand 35-shilling microscope to help him study blood and fibre samples from crime scenes. He also taught himself how to test for the presence of blood, making up his own reactive solutions in his bathroom at home, and worked out a way of deciphering erased handwriting.

Although Cuthbert's colleagues undoubtedly found his insights helpful, his immediate supervisors took a rather dimmer view of his activities, and he received reprimands on several occasions for paying more attention to his 'hobby' than to his official duties. In an attempt to curb his enthusiasm, he was transferred to the Criminal Records Office, where he would have less access to working police officers and therefore be less likely to cause trouble. The detective inspector to whom he reported was given strict instructions to keep a close eye on his new recruit and ensure that, during office hours at least, Cuthbert did only the work he was assigned and nothing else.

In response, Cuthbert moved his microscope and other items to the police section house where he lived and continued to assist colleagues while off duty. As word of his skills spread, he began helping out on crimes committed not just in London but all over the country. One case involved a document from Kent which had, allegedly, been tampered with. Cuthbert not only proved that the document had been altered, as the police had suspected, but also managed to recover the original writing, providing clear evidence of the motivation for it having been changed in the first place.

A few weeks later, he was ordered to report to the office of Sir Norman Kendall, assistant commissioner of the Met. A furious Sir Norman explained that he had received a message from Folkestone asking for Cuthbert to attend court to give

evidence on the document case. The request itself was out-
rageous, but what truly infuriated Sir Norman was that the
missive referred to Cuthbert as a member of 'the Metropolitan
Police Laboratory' – a facility that simply didn't exist.

Sir Norman flatly refused permission for Cuthbert to give
evidence and delivered another rebuke. The experience left
Cuthbert convinced he was about to be sacked at any moment.
He would have resigned on the spot but, as this was the early
1930s and so the time of the Great Depression, he knew he'd
struggle to find another job. Besides, any reference he received
from the Metropolitan Police was unlikely to be flattering.

A few weeks later, the court overruled Sir Norman and
issued Cuthbert with a subpoena, ordering him to attend and
give evidence. Despite being largely self-taught, it turned out
that Cuthbert had a firm grasp of the subject and, at the trial,
his clear and well-presented evidence was the key that helped
the prosecution secure a conviction. Cuthbert even received a
commendation for his work 'within the Metropolitan Police
Laboratory' from the judge, which eventually made its way to
the office of Viscount Hugh Montague Trenchard, the police
commissioner and the only Metropolitan Police officer more
senior than Sir Norman.

Intrigued by word of a facility within the Metropolitan Police
that he knew nothing about, Trenchard asked for a personal
tour of the lab. Cuthbert, now more fearful than ever of being
made to look a fool, quickly gathered up a loose collection of
odds and ends, along with his microscope, and spread them
out in a corner of the Met's photographic department. He also
borrowed a white laboratory coat from the local hospital in a
bid to look the part.

Despite his best efforts, it was still a pitiful display and
Trenchard was so appalled at what he saw during his tour that
he immediately forced through funding to establish a proper

laboratory as quickly as possible, promoting Cuthbert and asking him to supervise both the choice of equipment and its installation.

The new building – based at Hendon Police College – was opened on 10 April 1935 and contained lecture halls, dark-rooms and several well-appointed laboratories, complete with the latest microscopes, epidiascopes (a projector used to display images of opaque objects, such as drawings), centrifugal separators, X-ray machines and a reversion spectroscope.

The full-time staff included a pathologist, a chemist, a tech-nician, a clerk and a cleaner. Cuthbert himself was appointed police liaison officer. Over the course of the next few years, the small group analysed blood and semen stains, identified poisons and explosives, and examined tool marks, firearms and spent ammunition. The laboratory also initially had a teaching purpose and all new police constables were required to attend a course on forensic science there. This function was dropped after a few years, however, and the laboratory was expanded and moved first to the New Scotland Yard Norman Shaw building, then to Richbell Place in Holborn and finally to Lambeth, where it remains to this day.

The success of the London operation soon led the Home Office to sanction the development of forensic laboratories to provide services to police forces outside London. The initia-tive eventually saw seven independent laboratories established around the country, collectively known as the Home Office Forensic Science Service (HOFSS).

Cuthbert remained at the Metropolitan Police laboratory for the rest of his career and rose to the rank of superintendent before retiring. Ironically, considering the huge number of crimes he helped to solve, he left the force having failed to make a single arrest during his twenty-eight years as a police officer.

★

On the very first morning of my new job, I found myself sitting in a side office beside a quietly spoken, clean-shaven Scotsman as we filled in the multiple forms needed to get us on to the payroll. It was his first day, too, and we quickly bonded, becoming colleagues and forging a friendship that continues to this day.

Once the administrative tasks were done, our next step was to head off to Peel House to attend an induction course. Although we were civilians, we were still going to be part of the Metropolitan Police, a vast organization encompassing a wide variety of roles, ranging from dog handler or secretary to traffic warden or forensic scientist.

As a result of this sheer size and diversity, the powers that be had long ago decided that every civilian joining the Met should attend a two-week course in order to help them fully understand the culture of the organization and the various opportunities for advancement within it.

As my new friend and I diligently sat through the various presentations given by members of the HR department, union representatives, and so on, by the end of the first week we were struggling to see the relevance of it all. From my time as a student, I remembered that the Westminster Arms, a pub which was popular with the laboratory staff as, every Friday lunchtime, it put on a show in an upstairs room featuring 'exotic' dancers, was just around the corner from where the course was being held. The senior laboratory managers were also known to go there – but we all knew enough to remove or tuck in any necktie bearing the Metropolitan Police motif when we went.

We decided to go along, and ran straight into a number of our new colleagues, turning the whole experience into a rather enjoyable social occasion. Not wanting to walk out mid-show, we politely waited until the dancer had finished her performance before leaving and, as a result, were a few minutes late getting back to the course.

When we took our seats ahead of the next presentation, the man in charge of the induction course was less than amused. He informed us that we were no longer welcome on the course and that we should return to our boss and tell him we had been kicked off. With the ink barely dry on our employment contracts, we didn't relish the prospect of having to tell our new boss what had happened – especially as we hadn't even met the man at the time – but we had little choice.

In the event, he was every bit as furious as we expected him to be, though his anger was directed not at us but at the man in charge of the induction course. Not only were we allowed to start work right away without finishing the second week of the course, but also, from that day on, no one joining the MPFSL ever had to attend the course again.

At the scene of a crime, any stain believed to be blood (or any other body fluid that might have some bearing on the case) is subjected to a presumptive test before being sampled or collected. A positive result in a presumptive test does not confirm the presence of blood; it merely indicates that there is a likelihood that blood is present and that further testing should be carried out. If a presumptive test is negative, however, the presence of blood can be completely ruled out, so more time-consuming tests can be avoided.

In some cases, such as my encounter with the nearly headless body recounted earlier, there is little doubt that blood is present, but in cases where the evidence involves stains on clothing or other items which are weeks, months or even years old, such tests are essential.

As a result, one of the very first procedures I ever learned as a forensic scientist was the Kastle–Meyer test. Created at the start of the 1900s, the test relies on a chemical substance called phenolphthalein, once a widely popular laxative, which is put on a sample of bloodstain. After this, a small amount of hydrogen peroxide is added. If the enzyme peroxidase – a constituent of blood – is present, it causes the hydrogen peroxide to break down and release oxygen, which then reacts with the phenolphthalein, turning it dark pink.

The test itself is relatively straightforward. The scientist will fold a circular filter paper to obtain a quarter-circle section with a sharp point, scrape the point over the stain they wish to test, then open up the filter and add a drop of alcohol to make the stain soluble, followed by a drop of KM reagent, and one of hydrogen peroxide. A positive reaction, indicated by the colour change, should take only a second or two. (See page 5 of the first picture section.)

Although the KM test was – and continues to be – widely used, it has limitations, including the fact that it cannot distinguish between human and animal blood.

The colour change is caused by the oxygen released when the peroxide reacts with the phenolphthalein, but oxygen in the atmosphere means that all tests will usually produce a reaction within 30 seconds, so the speed at which the colour develops needs to be carefully observed.

The KM test will also produce a positive result when used on plant material such as grass stains, potatoes, horseradish and other substances such as ox-blood shoe polish. In some such cases, the colour produced can be less vivid, take a long time to develop or begin to form before the addition of hydrogen peroxide. So part of the skill required to use the test properly is having sufficient knowledge and experience to determine what is likely to be a genuine reaction and what is likely to be a false one. (The colour of the suspect stain can also give a clue!)

Over the months that followed, I learned how to perform a wide range of other presumptive and confirmatory tests and procedures, all under the direction and supervision of various reporting officers and other experienced scientists.

These included the acid phosphatase test for semen, the alpha-amylase test for saliva and the periodic acid–Schiff test for vaginal fluid. As with the Kastle-Meyer test, many substances can produce false positives. In the case of the test for semen, these include watermelon and cauliflower (which isn't usually too much of a problem, as they are rarely featured in sexual assaults or encountered alongside semen). However, both contraceptive creams and vaginal fluid itself can produce weak false-positive results.

Although I was chiefly going to be based in the biology department, all newcomers were required to spend time in all the other departments, too, to help them gain a better understanding of the work being done there.

Over the course of several months I worked alongside experienced analysts in the drugs, toxicology and alcohol sections of the chemistry division, tested for residues of flammable liquids in the fire investigation unit and learned how to match glass fragments by measuring the refractive index of samples.

In the documents section, I learned how to obtain and compare handwriting samples, to identify unique marks left on paper from individual typewriters and printers and a variety of techniques to find out whether a document is counterfeit or genuine.

In the photography unit, I learned about the specialized equipment used to record results of comparisons and analyses throughout the laboratory and also how to match negatives to particular cameras and identify the source of particular photographs. The Serious Crimes Unit was responsible for capturing pictures of fingerprints at the scenes of such crimes, sometimes using high-powered lasers and a range of chemicals to reveal the presence of fingerprints that would otherwise be invisible.

I was trained in the basics of forensic entomology – using insect larvae to determine the time of death – along with microspectrophotometry, to identify textile fibres and paint pigments and to use chemical methods to restore erased serial numbers on car engines.

In the fibres unit, I was taught how to collect, identify and compare natural, man-made and synthetic textile fibres, and also various techniques to extract dyes from individual textile fibres in order to identify those produced as part of the same batch-dying process.

The final stop on my laboratory tour was a brief stint in the firearms section, where I learned how to compare bullets to see if they had been fired from the same weapon and to identify firearm discharge residue particles with the aid of the laboratory's powerful scanning electron microscope.

By the time I returned to the biology department I was keenly aware that the often messy, hands-on approach to forensic science I had been using during my time as a research student was very much par for the course in the day-to-day lives of all the staff in the department.

Few of the materials, reagents and pieces of equipment we required were commercially available, so, frequently, if we needed something, we had to make it ourselves. When compared with the grey boxes and shiny lights of automated and computer-controlled modern DNA profiling, working in a forensic lab during the late 1970s was relatively primitive – a blend of DIY, science and using a home chemistry set in your kitchen.

When we needed anti-H (antiserum for type O blood cells) for serological grouping we had to grind up the seeds of a gorse bush and make our own; a type of bean was used to make a different agglutinin and, of course, for a short time at least, snail albumen glands were turned into an A1-blood type hae-moglutanin for bloodstains.

Anti-human antisera utilized to identify whether a blood stain was of human origin was taken from a small colony of rabbits that were kept in the animal house of the London Hospital. These creatures were periodically injected with human blood drawn from the laboratory staff; the rabbits made antibodies against the 'foreign' human blood and the rabbits' ears could be bled to harvest this anti-human antiserum.

Much of the equipment, too, was home-made. Compact humidity chambers did not exist, so we made do with sand-wich boxes and damp blotting paper. Textile-fibre dyes were extracted in our own home-made microscopic tubes con-structed for the purpose from capillary tubing with heat-sealed ends. We glued together pieces of 5mm Perspex to make electrophoresis tanks; the electrolyte 'wicks' we made from washing-up sponges. The tanks were to accommodate the electrophoretic plates used in the analysis of blood. The wicks were to carry the electrical current from the buffer wells in the tanks to the electrophoresis plates.

Perhaps one of the more bizarre creations occurred during the early development of DNA profiling. At one stage of the process it was necessary to rock a membrane gently in a bath of warm water. Appropriate equipment was not available at the time and so we knocked something up with a wire coat-hanger and a piece of railway track and a truck from a Thomas the Tank Engine set. The plastic (hence waterproof) track was glued to the bottom of a stationary water bath; the truck was placed on the track and connected to an orbital cam motor-drive with the wire coat-hanger. The bath was filled with warm water and the DNA membrane placed on the truck. The motor-drive switched on and – *voilà!* Such was the world of forensic science in those days.

Control samples of blood for each of the ABO types and the 15 or so other blood group systems in use at the time came from daily 'bleedings' of members of the lab staff. Those unlucky scientists with the misfortune to have inherited one of the rarer blood groups were in great demand every day and, as a result, were always a few shades paler than their colleagues.

As I slowly gained more experience, I began to take a more active role as an assistant in ongoing cases. Although most of my time was initially spent in the lab, I soon started to accompany reporting officers to actual crime scenes, helping to identify, test and collect samples of blood and other body fluids.

As I'd only been dealing with samples in the laboratory up until then, it was relatively easy to isolate myself from the crimes they related to. As my training progressed, however, I found myself coming ever closer to the real victims of the crimes which I was helping to investigate. I knew it was only a matter of time before I came face to face with my first dead crime victim.

In the event, I ended up spending more time looking at the inside rather than the outside of my first corpse, as the next part of my training towards becoming a fully fledged reporting officer was to attend an autopsy.

During the days leading up to it, I couldn't get it out of my mind. It wasn't just that I was going to see a dead body up close for the very first time, but also that it was going to be cut up. I had no idea what it was going to look like, or how it would smell. I'd heard stories about people passing out during autopsies and I would prefer not to be one of them. The only advice I was given in advance was in relation to the smell. It was suggested that I dab a few drops of perfume on to my shirt cuff so I could hold it up to my nose if the odours proved too much for me.

Deptford mortuary wasn't at all what I expected. Instead of the clean, clinical space you see on television shows, this was a converted tram shed with no air conditioning. Bodies were stored on wooden trestle tables in a circle around the edge of the main room and the body being dissected was in the centre on a ceramic table with a hollow in its surface so that the blood could be washed away.

It was, to be honest, a bit of a spooky setting. All the observers were gathered around the central table, and you could back up if you didn't like what you were seeing, but if you backed up too far, you'd brush up against one of the dead bodies lined up around the edges.

One thing I should mention: although the physical inspection of a dead body is often referred to as a 'post mortem', this translates as simply 'after death'. A 'post mortem examination' is more accurate but, within the world of forensic science and pathology, the usual term is autopsy.

There are two types of autopsy – forensic and clinical. The latter is frequently carried out to establish the exact cause of death in non-suspicious circumstances, while the former is a much more rigorous procedure when death is thought to be the result of murder or manslaughter, or there are other causes for suspicion.

To help newly recruited police officers and forensic biologists better to understand what is involved, pathologists carry out training autopsies, where they undertake a number of additional procedures for demonstration purposes.

The first autopsy I attended was being conducted by a middle-aged pathologist who seemed to take a certain amount of delight in shocking the audience as much as possible. This was long before smoking at work had been banned, and their party trick was to puff away on a fat cigar while carrying out the investigation and to use the open mouth of the corpse as an ashtray. It was clear that they enjoyed the fact that their actions made everyone else so uncomfortable.

In a normal forensic autopsy, a huge amount of time is taken up examining the outside of the body, as this is often where much of the evidence is to be found. If there are knife or needle marks, these need to be measured and photographed, and samples of hair and fibres have to be collected before the body is washed.

As we were merely watching a training session, however, the pathologist on this occasion was able to get on with the first incision right away. Rather than a scalpel, a large, incredibly sharp knife is used to slice a vertical line down the middle of the chest.

After death, the skin loses all its elasticity so, unlike the skin of a living person who is, for example, having an operation, that of a dead person needs to be drawn back in order to enable the next part of the procedure to take place. The knife is turned sideways and slipped beneath the base of the skin and the ribcage in order to loosen the skin so it can be pulled to one side.

Once the ribs are fully exposed, an instrument similar to a pair of bolt croppers is used to cut through them, along with the collarbone, so the entire front of the chest can be lifted away in a single piece, exposing the internal organs.

I had carried out lots of dissections as a student, but never on anything bigger than a dog-fish. Up until the time the first incision was made, I was very aware that I was looking at a dead body, but once it had been opened up it just became an incredibly interesting dissection.

The inside of a human body was not quite what I'd imagined. The organs were packed in extremely tightly; there wasn't a lot of space between them at all. Each organ was removed, the connections between it and the rest of the body severed, and examined under the bright lights, giving us all the chance to have a good look.

The liver was much larger than I thought it would be – it's huge – while the lungs seemed smaller and different in shape to what I'd anticipated. The smell was reminiscent of being at the butcher's and, much of the time, so was the procedure itself, especially when it came to separating the skin from the flesh. The body itself was quite fresh, so there was only minimal decomposition.

The only time I found myself feeling a bit queasy was when it came to removing the brain. First, an incision is made in the hairline, an arc that crosses the top of the head from behind one ear and over to the other. The knife is slipped into the gap between the skin and the skull in order to loosen it. The face is then peeled forward and turned inside out so that the top of the hairline is level with the chin. Depending on how much blood the body has inside it, the inside of the face is either white and pale or a red, bloody mess. The skull is usually clean and white: it almost looks as though the person is wearing some kind of fencing mask.

The pathologist (or, more usually, the mortuary assistant) then uses a small hand-held circular saw to cut a circle through the bone around the top of the skull. The saw is water-cooled and, as it cuts through the bone, the smell and noise are reminiscent of being at the dentist's. When this is complete, a small medical 'crowbar' is inserted into the gap and levered upwards to crack open the skull. It was at this moment that I felt my insides twinge. As it is normally sealed off from the rest of the body, opening the skull in this way causes air to rush in, creating an awful sucking noise that is somewhere between a crack and a pop.

Once the top of the skull is removed, the two large lobes of the brain itself are exposed. The pathologist cuts away at the base to allow the brain to come away from the skull completely, thus exposing the smaller organs that remain in a deeper pocket at the base of the skull.

It was the late 1970s, and I remember the Bee Gees were riding high in the charts. One of the leading fashions of the time – I even had one myself – was for men to wear white or cream three-piece suits with very wide lapels.

One of the DCs observing the autopsy was wearing just such an outfit. As the pathologist struggled to remove the hypothalamus with a large pair of forceps, we all moved in for a closer look. Suddenly, the organ came free and went flying across the room, landing directly on the left lapel of the DC with a wet, slapping sound.

The man looked down at his lapel, saw the bloody grey lump of brain matter he was now wearing and went into a dead faint. It was as if someone had just turned him off.

He came round soon enough, though, and once the autopsy was over, everyone headed outside for a breath of fresh air. I was proud of myself for having come through the experience relatively unscathed.

A few weeks later I attended a second training autopsy, this time at a more modern mortuary. It was performed by Dr Iain West, considered to be one of the top pathologists in the country and the man later responsible for carrying out autopsy examinations on the likes of media tycoon Robert Maxwell, television personality Jill Dando and police officer Yvonne Fletcher, who had been shot dead outside the Iranian embassy in 1984.

Like the pathologist before him, Dr West seemed to delight in making trainees wince as much as possible during the course of the procedure. After carrying out the usual examinations, he asked the all-male team of observers to gather around more closely so that they could observe exactly what he did next.

He hovered over the man's pelvic region and, suddenly, without any warning, he flashed his knife and cut the man's penis in two, lengthways. Before anyone in the audience could catch their breath, he then chopped the man's testicles into thin slices as if they were a pair of boiled eggs.

It was all done with a degree of flair and was of limited scientific value, but the fact that I still remember it forty years on shows that it had the desired, shocking, effect.

In the case of death by natural causes, the body can remain relatively unchanged for the first couple of days, and most people rarely see bodies that are any older unless they have been embalmed. This process, which replaces the blood and other body fluids with chemical preservatives, slows the growth of the bacteria responsible for decomposition.

However, many of those who meet their deaths by stabbing, blunt trauma, gunshot, drowning, fire or strangulation are not discovered for days, or even weeks, and during this time the body can undergo dramatic changes, far beyond the imaginations of film and TV special-effects teams and make-up artists.

Within the space of a few days, the skin becomes discoloured, first turning green, then purple and, finally, black. At the same time, bacteria inside the body begin to liquefy the internal organs, creating a foul-smelling gas that builds up, causing the abdomen to swell and bloat. By this stage, visual identification of an individual may be almost impossible, as physical changes mean that thin white males can easily be mistaken for obese Afro-Caribbean males.

As days turn into weeks, the skin separates and begins to slough off. Hair, teeth and nails fall out. As more gas builds up inside, the tongue is forced out of the mouth, the eyes bulge, fluids leak out of all the orifices and the stomach threatens to burst open – which it often does.

The pace and progress of decomposition is heavily dependent on environmental conditions. In dry conditions, a body can mummify, and the skin becomes dark, dry and leathery. In cold, wet environments, the fat beneath the skin can saponify (turn into soap), helping to preserve the body in relatively good condition for weeks.

If the face is still intact, the expressions can vary from serene and peaceful to shock and horror. Eyes and mouths are often open and refuse to stay shut. Within days – less if the ambient temperature is warm – body fats begin to break down. Flies are attracted to the liquefying fat and begin to lay their eggs in the orifices of the body. As maggots develop, they first appear in the corners of the eyes, inside the nostrils and the mouth, as well as inside any open wounds.

Heat from a fire can cause bones to break and skin to split. Joints contract, bending the arms and legs and leaving a body looking as though it has been in the middle of a boxing match, fighting for its life when the end came.

It was clear that, to be an effective forensic scientist – especially a reporting officer – I would need to develop a strong stomach. I would have to disassociate the fact that what I was seeing in front of me used to be a living person and focus all my attention on the job in hand. The dead cannot say how they died, so it was up to me to help figure it out on their behalf.

4

AT EVERY CRIME SCENE, INVESTIGATORS BASE THEIR ACTIONS on a principle which, despite first being formulated more than a century ago, remains as valid today as ever. It was devised by French police scientist Edmond Locard in 1910 and states that no two objects can be in contact with one another without evidence of that contact being left behind on both.

In practical terms, this means that when a burglar, killer or rapist commits a crime, they not only leave behind evidence of their own presence but take something from the scene away with them. Most of the time, the principle is reduced to a single, succinct sentence: every contact leaves a trace.

Dr Paul L. Kirk, a pioneering forensic scientist in America in the years following the Second World War, put it rather more elegantly in his 1953 book, *Crime Investigation* – so elegantly, in fact, that today his words are routinely mistaken for those of Locard himself:

Wherever he steps, whatever he touches, whatever he leaves, even unconsciously, will serve as a silent witness against him.

Not only his fingerprints or his footprints, but his hair, the fibres from his clothes, the glass he breaks, the tool mark he leaves, the paint he scratches, the blood or semen he deposits or collects. All of these and more bear mute witness against him. This is evidence that does not forget. It is not confused by the excitement of the moment. It is not absent because human witnesses are. It is factual evidence. Physical evidence cannot be wrong, it cannot perjure itself, it cannot be wholly absent. Only human failure to find it, study and understand it, can diminish its value.

Except in the case of major crimes such as a murder, forensic scientists don't usually visit the scenes of the crimes they help to investigate. More often than not, the first they see of a particular investigation is when a set of swabs, samples or exhibits arrives on their workbench.

In such cases, scenes of crimes officers, some of whom will be police officers themselves, some civilians, will have collected the samples and sent them in.

Having spent months assisting on cases, carrying out dozens of examinations and getting to know all the various sections of the lab, there was one last hoop I had to pass through before I would be allowed to take on cases of my own as a trained reporting officer – to experience a crime scene through a SOCO's eyes. To this end, I booked into a guest house in Brighton, Sussex, for two weeks and made my way to the local police station in order to accompany the two-man SOCO team to each and every scene they were called on to attend.

A good SOCO acts as the eyes and ears of the scientists back at the lab, recovering material including blood, fingerprints, shoe marks, paint, fibres, glass and anything else that could be evidence. Small objects and particles of materials such as glass are placed in evidence bags, and blood and other body fluids are swabbed, while fibres, fingerprints and shoe marks are 'lifted' using adhesive tape. In the case of fingerprints, this involves making the mark visible by dusting it with special powder, pressing down on the powdered mark with a length of adhesive tape and then carefully 'lifting' away a perfect copy of the ridge pattern.

Some of my colleagues at the laboratory sometimes complained about the quality of the samples they received from the SOCOs, and part of the reason reporting officers were encouraged to spend time with them was to get a first-hand understanding of the day-to-day difficulties they face.

I was eager to gain as much hands-on experience as possible but, as each home or commercial premises we went to was a live crime scene, there was no way in the world my companions were going risk evidence being destroyed. Instead, I patiently observed as the pair of officers lifted prints and collected samples. It was only once they had covered the key areas of the scene that I'd be allowed to have a go myself.

It didn't take me long to realize that dusting for fingerprints is nowhere near as easy as it might appear on television. Using the right amount of powder and a light enough stroke with the brush is only half the battle. As most fingerprints are invisible, you need to know the best places to look in the first place, unless you want to spend your time dusting every inch of a property. And lifting a print takes almost as much skill as powdering it, as putting pressure in the wrong place can easily lead to a smear which renders the print useless.

By the end of my two weeks I had a huge amount of respect for the SOCOs. In the years that followed, commercial concerns meant that future reporting officers would miss out on the chance to work alongside them. I would sometimes hear these colleagues complain about the quality of a shoe print or photograph and question the skill of the collector, but having myself tried to lift shoe marks while halfway up a wall in a howling gale, or make casts of tyre tracks before a rainstorm came in and washed all the evidence away, I was more aware than most that circumstances were not always ideal.

Soon after getting back from my temporary secondment to Sussex, I found myself heading straight back there in order to start work on my very first murder case as a reporting officer.

Although it was exciting knowing that, whatever I found, I'd have to present evidence about it in court, the case itself initially looked straightforward and it seemed that there was every chance it might turn out to be something of a damp squib. There was no crime scene, no murder weapon and no blood pattern evidence. There wasn't even a body for me to view, as the initial autopsy had ruled the death accidental, and it was only when new information came to light that two subsequent autopsies found the death to be deliberate. If she had known how it was to turn out, I doubt my boss would even have allocated the case to me, as I was such a new reporting officer.

The case involved Colin Wallace, an information officer with Arun district council, who had been accused of murdering his friend, antiques dealer Jonathan Lewis, after forming a close bond with Lewis's wife, Jane – a work colleague – a bond that was on the verge of boiling over into a full-blown affair. Lewis's body had been pulled out of the River Arun, around three miles downstream from Arundel, on 8 August 1980. He had been reported missing four days earlier after failing to turn up at a restaurant where he was due to have dinner with his wife and a few friends, including Wallace and Wallace's wife.

According to the police, the murder had been particularly callous. They alleged that Wallace had secretly met up with Lewis earlier that same evening and, after getting into an argument, killed him instantly with a single blow from the heel of his hand to the base of Lewis's nose. Wallace was a former captain in the British Army and had served with the Special Forces, so he had all the necessary training in unarmed combat to be able to make such a deadly strike.

Then, it was alleged, Wallace had stuffed the body into the boot of his car and driven to the dinner party, playing along with the other guests – including Lewis's wife – as they became increasingly concerned about Lewis's whereabouts. At some point Wallace complained of stomach pains and left the dinner to go home, as he said, to take some medicine. According to the police, however, he had taken this opportunity to dump the body in the river.

Wallace's car, a distinctive white Austin Princess garishly decorated with Union flags, was duly brought to the Lambeth laboratory on a low-loader lorry for my examination. Subsequently, it was collected by Sussex police. A couple of weeks later, after a further SOCO examination, suspicious stains were found in the boot of the car and I was called down to Littlehampton police station to examine it and collect samples for grouping which could then be compared to samples taken from the victim and the suspect (see page 8 of the first picture section).

As is the usual procedure with sampling at crime scenes, I first tested the various stains in the boot of the car with KM reagent. Where a result was positive, I circled the stained area with a wax pencil. Using a sterile cotton medical swab, moistened with distilled water, I swabbed up the bloodstain and sealed the swab in a labelled tube. A reference number was given to each sample and its position marked on the car.

Back in the laboratory, I cut pieces from the swabs and made extracts from the unidentified bloodstains in the wells of a glass microscope slide. I used tiny single threads of clean, boiled white cotton sheet to soak up the individual blood extracts in order to carry out blood group analysis, including control samples of known blood group variants taken from laboratory staff.

While ABO typing using whole, liquid blood taken directly from a suspect is pretty straightforward, the method doesn't work with stains, as the red blood cells rupture as the stain dries, leaving no cells to agglutinate directly. However, the antigens present on the surfaces of the cells remain intact and can be identified using a process known as the absorption-elution technique.

I'd start by using a wax pencil to divide a small sheet of acetate into a grid of 30 evenly sized squares arranged in six columns and five rows. The left-most column and top row of squares were reserved for labelling, while the others would be used for samples. I'd then cut a series of threads, each about half a centimetre long, from the bloodstained fabric I was trying to analyse and use a drop of glue on the end of a toothpick to attach each thread to a square. Next, I'd add a drop of anti-A antisera to the first column of threads, anti-B to the second column and our own 'home-made' antisera for O cells (known as anti-H) to the third column. The two remaining columns were control samples, labelled OA and OB, to check the results against. I added serum from a group O individual (containing both anti-A and anti-B) to these columns.

After this, I'd place the acetate inside a humidity chamber – as I mentioned before, these could be really makeshift: a plastic sandwich box with a piece of wet blotting paper stuck to the inside of the lid – and leave it for a while to allow any matching antibodies to be absorbed by and bind to their equivalent cell-surface antigens on the threads. I would then wash off the excess antisera, blot the acetate sheet dry and add a red blood cell solution to the threads: A cells to the first column, B to the second, and O to the third. A and B cells were added to the OA and OB control columns respectively.

The next stage of the procedure was to place the sheet in a humidity chamber in a warm oven, where, in a process known as elution, any absorbed antibodies are released from the threads and mix with the test blood cell solutions. I'd then be able to examine the acetate sheet under a low-power microscope to look for signs of clumping in any of the red blood cell solutions.

Agglutination is scored on a scale that runs from 0 (no agglutination) to 4 (complete agglutination). If agglutination is obtained only with A cells, then A antibodies have been absorbed and then eluted, which means that the original bloodstain has A antigens and is therefore from a person of blood type A. Similarly, agglutination only with B cells means blood type B; with only O cells, blood type O; and with both A and B cells, blood type AB.

If the body fluid is not blood but semen or saliva, then a different technique has to be used, as there will be no blood cells to agglutinate. Instead, I'd be looking for the presence of free ABH substances in the semen or saliva. In the 86 per cent of people who are 'secretors', the same ABH substances present in their blood can also be found in their other body fluids.

The test for these secretions is similar to the absorption-elution technique, except that the corresponding antiserum binds directly to the ABH substance and will therefore not be available to agglutinate the appropriate test blood cell solution. In this procedure, I'd be looking for an absence of cell clumps to indicate a match. Because the reaction is prevented by the corresponding antiserum, this process is known as the absorption–inhibition technique.

The blood from the boot of Wallace's car turned out to be type O – the same as that of the victim, Jonathan Lewis – but with

45 per cent of the UK population sharing the same group, it was impossible for me to draw any sort of inference. There were, however, at least a dozen or so additional analyses that could identify particular proteins or enzymes present in the blood and in doing so help to individualize a blood type.

Each individual test requires a cotton thread soaked in blood from the crime scene. With dozens of threads needed to carry out the full range of analysis and each thread being used just once, it's easy to see why blood grouping can be effective only when stains are sufficiently large.

Although they were not able to pinpoint individuals the way modern DNA profiling can, the results of a full battery of these tests could still be pretty impressive. In the same way that blood is subdivided into different groups, so enzymes and proteins have different variants.

For example, adenylate kinase (AK) comes in three forms, while phosophoglucomutase (PGM) has 10 forms. Extensive population studies have determined the percentage of people who have each of the various forms and, because the variants are independent of each other – that is to say, if an individual is ABO group A, they can still be any one of the 10 PGM variants – you can multiply the percentages for the occurrence of each variant together in an individual.

So, as we have seen, individuals with type A blood make up around 42 per cent of the UK population. If one of those individuals also has the type 2-1 variant of the protein AK (8 per cent of the UK population) and the type 1+ 1- variant of the enzyme PGM (17 per cent of the UK population), the occurrence of those three blood group variants together occurs in only half of 1 per cent of the population (0.42 x 0.08 x 0.17 = 0.005).

Analysis of these enzymes requires the use of a technique called electrophoresis – a process that allows biological molecules to be categorized according to their size. Using an electric field, molecules are made to move through a specially prepared gel. The larger molecules move more slowly, while the smaller molecules move faster. Different-sized molecules therefore form distinct bands on the gel, which can then be photographed. The position of these bands shows which particular variants of the enzymes are present.

Of the three variants of the AK protein, 1, 2 and 2-1, 2 is uncommon, 2-1 less so, and 1 is the most common, making up 80 per cent of the population. The analysis I had performed on samples in the Wallace case had come back with a plate showing images that made the blood unlikely to belong to group 1, but it wasn't possible to say whether it was a 2 or a 2-1. The bands always appear in specific places and, though there was a very faint image in the exact place where the 2 band was located, it was nowhere near strong enough for me to say it was a positive match (see page 6 of the first picture section).

In TV dramas, virtually everything that is tested provides a clear positive or negative result but, in real life, far more often than not, results fall into a grey area. Somewhat disappointingly for my first court case as a reporting officer, that was the situation I was facing. It simply wasn't possible to say with any degree of certainty whether the blood belonged to the victim, to the suspect or to any other particular individual.

I tried having the original electrophoretic plates 'overcooked' in an attempt to develop the bands more, but that didn't help. I then sent the original photographs and negatives away to the Police Scientific Development Branch (PSDB) – the Home Office department set up to provide advice, support, technological and operational capability to the criminal justice system – as I hoped their research might be of help.

At PSDB the images were redeveloped with a variety of colours and made negative rather than positive in an attempt

to enhance them. The result was still the same – inconclusive
– and I reported it as such in my statement.

Despite this, I would still have to go to court and report
my findings, but that would take place several months in the
future. In the meantime, I had plenty of other cases to occupy
my time.

The following week, I was called out to the scene of a woman's
death. It looked originally as though her death had been of
natural causes – but then a series of apparent bloodstains were
discovered near by and a group of puncture wounds detected on
the deceased. It was my first time at a scene with a body *in situ*.

Before I had even entered the small flat on the second floor
of a council block in Hackney, I noticed the smell of frying
bacon. It was just coming up to the early hours of Sunday
morning and I hadn't yet had breakfast. I immediately began
salivating, wondering if the husband of the deceased woman
was serving up a treat for the investigating officers and hoping
there were still a few rashers left.

The moment I stepped into the lounge, I could see my
mistake. The flat belonged to an elderly couple, and when his
wife had died the husband had initially been alarmed by the
fall in her body temperature. In order to keep her warm, he
had moved her closer to the gas fire and turned it up to full.

It wasn't bacon that was cooking. It was his wife.

The bloodstains turned out to be old and of no significance.
As for the puncture wounds, the husband had been repeatedly
prodding his wife's body with a fork in an effort to rouse her.
The death was ultimately ruled to be the result of natural
causes, after all, but it was months before I was able to go anywhere near a bacon sandwich.

I was rapidly gaining lots of confirmation that a strong
stomach was a prerequisite for working as a forensic scientist,

but still fate seemed determined to do everything possible to ram the point home.

A month later I found myself at an especially grisly scene. A man living in the top flat of a terraced conversion had been murdered, and his killer had stuffed his body in the airing cupboard before making his escape.

The warm temperature had accelerated the decomposition process, so the first the neighbours downstairs knew anything was wrong was when dozens of live maggots started falling out of the light fitting in their living room and the stench of rotting, decaying flesh began to fill the common hallway.

By the time the investigation team arrived, the maggots had literally made a meal of things. The smell of decomposition inside the flat itself was almost overpowering, and it got stronger as we moved closer to the source. As the door to the airing cupboard was opened, a slimy mass of bones and half-eaten flesh – along with thousands of fat, well-fed maggots – oozed out on to the floor.

Soon afterwards, I travelled to a scene where a woman and her boyfriend had been murdered by her former lover. The pair had been fully decapitated with an axe while in bed and their bodies had lain undiscovered for more than a week. As I entered the property I could hear a mechanical droning noise like a piece of industrial machinery. I couldn't work out what was making it until I opened the door to the room where the bodies were and saw that the air was black with blowflies. The window had been left slightly open, so the flies were able to find their way inside, past the net curtain, and lay their eggs on the bodies but, once the larvae had matured and turned into flies themselves, they had been unable to find their way out.

The flies were absolutely everywhere and every time I opened my mouth to speak I would swallow one or two. Considering what they had been feasting on for the past few days, it was a stomach-churning experience, to say the very least. And this was in the days before protective face masks were worn as a matter of routine.

In preparation for becoming a reporting officer and my inevitable first appearance in the witness box, I had spent many days going along to various courts to watch my colleagues giving evidence. I was naturally anxious at the prospect of appearing in the witness box, but experienced colleagues assured me that scientists were rarely challenged on their evidence and that, so long as I had all my science straight, I wouldn't have anything to worry about.

This certainly seemed to be the case at the first few trials I observed. The forensic scientist would be sworn in, present their findings to the prosecution, clarify a few points for the defence during cross-examination, be thanked by the judge and then allowed to leave. The attitude of the courts was that we were experts in our fields, knew exactly what we were talking about and were capable of giving independent, unbiased opinions which jurors could treat as gospel.

None of my colleagues had ever admitted to a particularly hard time in court, so there was no reason to expect things would be any different during my first appearance as an expert witness. I received plenty of advice about how to conduct myself, the importance of speaking clearly and not answering back to the judge, but little else. Unfortunately for me, my career had begun just around the time the attitude towards expert witnesses was starting to change.

The catalyst for this switch was the fallout resulting from two incidents in which senior reporting officers from the Home Office Forensic Science Service (HOFSS) had presented evidence that later turned out to be badly flawed, leading to miscarriages of justice.

The first of these incidents dated back to November 1974, when two bombs exploded in two separate pubs in central Birmingham, killing 21 and injuring 182 more. The devices were thought to have been laid by the Provisional IRA, who, it was said, had hoped for an even higher death toll: a third bomb was later found outside a bank, but it had failed to detonate.

The police moved swiftly and by the end of the evening had made a series of arrests. The main suspects were six men, mostly Belfast-born Roman Catholics, all of whom had been living in Birmingham for some time. Five of the men had left New Street station just before the bombings to travel to Belfast to attend the funeral of an IRA man who had accidentally killed himself two weeks earlier while attempting to plant a bomb in Coventry.

Stopped by Special Branch officers who were monitoring anyone travelling to Belfast in the aftermath of the bombings, the men lied about their reasons for heading to Northern Ireland, were arrested and taken into custody.

Forensic scientist Dr Frank Skuse was called in and performed the Griess test — the standard test to detect traces of nitroglycerine — on swabs taken from the fingers of all six men. Two of the men, William Power and Patrick Hill, returned positive results.

Wholly convinced they had the right men, West Midlands Serious Crime Squad took over the case and began turning up the pressure. Within days, four of the six men had confessed their guilt.

At the trial, which began in June 1975, Dr Skuse told the jury that the results of the Griess test made him 99 per cent positive that Hill and Power had handled explosives prior to the bombings. He remained convinced despite the fact that all the other men had tested negative and that later, more advanced tests had not supported the initial test results.

The scientist called as an expert witness by the defence, Dr Hugh Black, said he did not believe the men had been in contact with explosives, but it was the evidence of Skuse that the jury preferred. It found the six men guilty of murder. They were each given 21 life sentences.

Over the years that followed, doubts started to emerge about the techniques used by Skuse and his flawed interpretation of the Griess test. It would later emerge that, because of the way he had set up the test, he would have received a positive reaction for all manner of things, including, crucially, the nitrocellulose coating of the playing cards which the defence claimed the two men had been passing the time with during their train journey from Birmingham to Morecambe where they would catch the ferry to Belfast.

The second incident involved evidence provided by Dr Alan Clift, a senior HOFSS analyst, in a rape case. He failed to inform the court that body fluids he had identified as coming from the attacker could in fact have come from the victim. When asked why he had not mentioned this crucial fact, he simply replied, 'No one asked me.' Dr Clift had refused to allow his written statements for the courts to be peer-reviewed by his colleagues.

In the months that followed, forensic scientists appearing in court had their methods and expertise questioned more and more. Cross-examination became less routine, often beginning with an attempt to establish whether the scientific expert witness knew or had any link to either Clift or Skuse, in which case an attempt to discredit their findings would invariably be made.

A month or so before the start of the trial of Colin Wallace, the former army officer accused of murdering the husband of a woman he was on the verge of having an affair with, I attended a case conference with the prosecution barrister to explain the findings I had detailed in my statement. Eager to be as helpful as possible, I told him why I had reached my decision and then showed him the images of the enhanced photographic plates.

The barrister then asked if I would amend my statement to expand on my view and to include the fact that the result of the AK blood analysis could have been either a 2 or a 2-1, even though that wasn't what I was saying at all. I tried to explain that the whole point was that I couldn't say this with any certainty, but he insisted I produce a new statement anyway.

The key issue was that Lewis, the murder victim, was a 2-1 AK group, while Wallace, the suspect, was a 1. If the blood was more likely to be 2-1 than anything else, it pointed more strongly towards Wallace's guilt.

I knew that, as far as the court and the defence were concerned, writing the new statement would make it look as though I had suddenly changed my mind at the eleventh hour, but I hadn't. I was as adamant as ever that the results were inconclusive.

My evidence in chief was pretty straightforward and took only a couple of hours. However, any hope that I'd get the same easy time under cross-examination as my colleagues had in the past quickly faded away when the defence barrister hit me with a series of tough questions, using the new statement as the basis for questioning my results and pressing me to accept that the blood was most likely a 2 and therefore did not belong to the victim.

The problem was that the 2 group was, statistically, much less likely to occur. In an effort to get this across to the jury, I came up with what I thought was rather a good analogy.

If a ship has three masts of different heights and sails over the horizon, an observer will initially see all three masts, then, as the ship moves away, see only two and, finally, only one.

Of course, it's still a three-masted ship; it just doesn't look like one any more. Now, if ships with one mast are extremely rare but ships with two and three masts are far more common, it means that if someone asks you to look at the ship just before it vanishes and there is only one mast visible and then asks you to say what you think the vessel is, you'll say that, although it looks like a one-masted ship, it probably has either two or three masts.

The cross-examination continued for two full days. The questioning was relentless, and in the end I was almost worn out. Everyone was getting frustrated and the judge simply wanted a straight answer one way or another. He would not accept my claim that the result was simply inconclusive as a consequence of my having written a supplementary statement.

They pushed and pushed, so at last I said that, if I was literally being forced to make a decision other than that it was not a 1, it was more likely to be a 2-1 than a 2, because it is far more common.

The judge was furious and accused me of having changed my mind about my evidence. The defence had done a great job of making it look as though I simply didn't know what I was talking about.

The moment the defence heard my definitive answer, the barrister sat down. I asked the judge if I could clarify my last point, but he told me that I had finished in the witness box and was free to leave.

(Despite my lack of scientific evidence against him, Wallace was ultimately found guilty of manslaughter and sentenced to ten years' imprisonment. In 1996, a decade after he was released from prison, his conviction was finally quashed and he was fully exonerated based on new conclusive forensic findings.)

My first court case had been a true baptism of fire, and I

could not have been more relieved when it was finally over. I felt as though I'd gone 15 rounds in a heavyweight boxing match and was physically and mentally exhausted.

I learned that day that you are always a hostage to fortune whenever you appear in court. Although the science is important and has to be right, it's not the be all and end all. The bat and the ball belong to the barristers. They have the script and they are the directors. Witnesses – experts or not – are just bit players. If the judge has had enough and no longer wants you to speak, you have to stop. If you try to clarify something or get your point across when the defence has finished its cross-examination, you risk being sent to prison for contempt.

It was an experience that affected me deeply in terms of my philosophy towards the job. While I absolutely wanted to remain a reporting officer and to continue giving evidence in future court cases, I knew that I'd always have to be extremely careful about expressing opinions or even hinting at possibilities that were not fully supported by the evidence.

That said, I never regretted my decision, and if I found myself in the same situation today, my evidence would be exactly the same. It's all too easy to try to avoid an aggressive cross-examination by reporting less than perfect results as 'inconclusive' and hope that you are not challenged on them. At the same time, it's also all too easy to end up being given a brutal time in the witness box if you have dumbed down results that might otherwise have benefited the defence.

5

IN THE SUMMER OF 1985, AROUND THE SAME TIME I HAD personally come to terms with the vagaries of the criminal justice system with regard to forensic experts giving evidence in court, Jayne Scott, a PE teacher at a secondary school in Slough, was consumed by fear that she was about to become the victim of a massive miscarriage of justice.

She had been accused of carrying out a savage hammer attack on her school's deputy head, Susan Craker, leaving her with such severe brain damage that she would have to spend the rest of her life in a wheelchair. Scott insisted that she was completely innocent, that an unknown intruder had been responsible for the assault and that Craker's horrific injuries had left her confused and clouded her memory.

The day of the attack had started innocently enough, with Scott, Craker and another woman, Deborah Fox, having break-fast together in the kitchen of Craker's new house in Barnet, north London, having all stayed over the previous evening. After they had cleared away the dishes, Fox had popped out to the shops and Scott had run a bath, while Craker sat cross-legged on the floor by the patio doors in the living room, reading a newspaper and drinking a cup of tea.

As she got into the bath, Scott recalled hearing a noise from the rear garden, but this stopped and was replaced by the sound of a slamming door, which seemed to come from inside the house, followed by a groaning, moaning sound.

Scott wrapped a towel around herself and made her way downstairs and into the lounge, which seemed to be where the noise was coming from.

'I saw Sue lying on her back. Her head, arms and legs were moving. Her legs were bent, her head was over to one side. She was making a groaning noise,' she told the investigating officers.

'By her right hand was a hammer. She was covered in blood, her face was covered in blood. I lifted the hammer out of the way to get closer to her. Blood was flowing out from behind her ear. I touched where it was coming out and then I just panicked. I felt dizzy and sick.

'The blood on my hand was all warm. I am not very good with blood at all. I just wanted to get away from it. I ran back upstairs to the bathroom. The blood felt hot and sticky. I leaned against the bath, feeling dizzy. I steadied myself with my left hand, while I shook my right hand into the bath. I switched on the taps and used my nightie to wipe the blood.'

Scott then heard the front door open. 'I shot straight down. Debbie had just come in. I was shaking. I said: "It's Sue, it's Sue." She asked where she was and I pointed to the lounge. She went in. I did not go back into the room. Debbie was quite calm and went to phone for an ambulance. It hadn't crossed my mind to call one. I just wanted to get the blood off my hands.'

Sue Craker was rushed to hospital, where she slipped in and out of consciousness. As Detective Constable Joseph McGahren tried to conduct an interview, it became clear that it was taking

Craker a huge effort of will to be able to say anything at all. 'Scott . . . hammer . . . struck' was virtually all she could manage.

DC McGahren attempted to clarify exactly what he was being told, asking Craker whether she had actually seen Jayne Scott hitting her. 'Hit first,' came the faltering reply, 'then Jayne.' Soon afterwards, Craker slipped into a coma.

Having got everything he could from the victim, DC McGahren returned to the house to speak to Scott and Fox.

From Scott's point of view, it was obvious what had happened. Craker had been alone in the living room when an attacker had come in through the patio doors. He had attacked her with the hammer and then fled back out through the garden when he heard Scott coming down the stairs.

Based on what Craker had told him, DC McGahren suggested that there was no attacker and that Scott herself had been responsible. 'I couldn't hit anyone with my hand, let alone a hammer,' Scott told him.

If, she insisted, Craker had seen her moving the hammer out of the way and felt her trying to tend to her wounds, she might easily have been confused about what was happening. 'I have no reason to fabricate,' Scott continued. 'I am not the person responsible. And I never tell lies.'

The corner of the lounge where the attack on Craker had taken place was covered in blood spatter, so the senior investigating officer made an immediate decision to put a call in to the laboratory to get a bloodstain pattern expert out to the scene.

By the time I arrived, suspicion was still being directed at Scott, but it was far from clear-cut, as the idea of an unknown intruder did seem to fit in with elements of the evidence. The patio doors were open when the ambulance crew arrived – Fox recalled them being shut when she left to go to the shops – and there were obvious spots of blood on the paving outside.

It was also known that at least two separate serial rapists were active in north London at the time, one of whom had struck in nearby Hadley Wood. In one incident, the rapist had incapacitated his victim with a hammer.

It was also clear that whoever had carried out the attack would have been covered in Craker's blood. The third woman in the group, Deborah Fox, had been out of the house for a maximum of seven minutes. She had not seen any blood on Scott and neither had a neighbour who arrived on the scene a few moments later, after the alarm had been raised.

There are two sides to every story and, in many criminal cases, the job of the forensic scientist is to ascertain which of the opposing scenarios – the one presented by the defence or the one presented by the prosecution – is best supported by the evidence.

If you enter a room that contains multiple blood splashes and distributions and are asked to describe what went on, you would find yourself embroiled in a research project that would likely consume you for weeks. Thankfully, what usually happens is that there are several possible scenarios offered for what has taken place, which allows you to focus your attention on which, in your opinion, is the more likely. With that in mind, I set about examining the scene.

The first in–depth use of bloodstain pattern analysis in a criminal context took place in America in the mid-1960s, when Dr Paul L. Kirk testified for the defence in the case of Dr Sam Sheppard during his retrial for the murder of his wife.

Sheppard had initially been convicted and imprisoned for the murder but had always claimed he had arrived home to find his wife dead and then been confronted by a bushy-haired man whom he believed to be the real killer. Although Sheppard never dramatically escaped from custody, his case would go on to inspire the popular television series and later film *The Fugitive*, in which the main character, Dr Richard Kimble, finds himself in a similar situation.

During the retrial, Kirk produced the results of a series of blood spatter experiments he had carried out, showing that whoever had killed Mrs Sheppard would have been covered in blood and would have to have been left-handed.

When police arrived at the scene of the murder, Sheppard, who was right-handed and not ambidextrous, had only a single bloodstain on him, on the right knee of his trousers, allegedly caused when he stood next to the bed to take his wife's pulse in the hope that she was still alive.

Kirk also demonstrated that the tracks left by the murder weapon indicated that it was likely to be a heavy torch, not the specialized medical instrument which the prosecution had originally claimed had been used. Second time around, Sheppard was acquitted of all charges.

Since that time, blood spatter analysis has become an increasingly important part of forensic science. Though by this point I had worked in a number of disciplines, blood patterns were by far my favourite and I was glad to have the opportunity to work another case and increase my level of experience. Even outside working hours I was reading books and studying articles and papers about the interpretation of bloodstains, and carrying out a few small experiments to see how drops of liquid – sometimes blood, sometimes not – behaved in different scenarios. I hadn't been recruited by the laboratory to specialize in this area, and expertise wasn't a requirement, but I found it fascinating and wanted to learn as much as I could.

When I arrived at the house in Barnet it was clear that the assault had taken place in the lounge and nowhere else, but I checked the remainder of the house anyway, as I wanted to be ready for any question the defence might throw at me.

The weapon had been identified and the blood patterns in the lounge were exactly what you would expect. They indicated upwards of five blows at a low level; the victim had not moved during the attack. It was clear that there had not been a struggle, that it wasn't a matter of self-defence – it was a question of someone bludgeoning the victim who was already down and out after the first blow.

Most of the blood was on the wall, with some splashes on a wicker chair in the corner, but what was most interesting was the blood on the patio door. Some blood had hit the paving stones outside, some the glass on the inside of the door and other blood splashes were on the lower runner on which the door moved. I could see where the blood had gone; what I needed to know is where it had come from. (See page 4 of the second picture section.)

Blood conforms to the usual properties of fluid dynamics and so forms spherical droplets as it flies through the air. This has the advantage that, when it does eventually hit a solid surface, the resultant stain can be measured and the trajectory of the blood droplet calculated. The width of the bloodstain represents the diameter of the droplet and its length is determined by the angle of impact. The relationship between the width and the length of the bloodstain is the sine of the angle of incidence $\sin \theta = \frac{W}{L}$. (See diagram on page 74.)

Using a powerful hand monocular lens with an accurate measuring graticule, I set about measuring the width and length of the most pertinent bloodstains. Previous experiments in the laboratory and experience had taught me which of the stains were likely to give me the most accurate results, so I made these my priority.

If you watch any of those slickly produced American crime

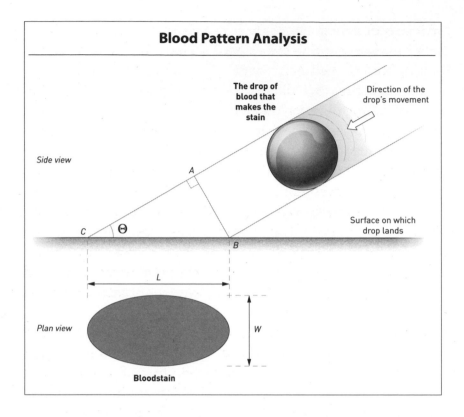

dramas, you'll see forensic scientists re-creating blood and bullet trajectories using lasers or high-tech computer programs, but in 1985 the system in use by the MPFSL involved a ball of string, a roll of electrical adhesive tape and a school protractor.

I proceeded to affix one end of the string to a bloodstain, used the protractor to measure the angle of incidence, then stuck the other end of the string to a convenient wall so that it indicated the correct trajectory. I then moved on to the next bloodstain of interest and repeated the process. The most difficult part of the whole procedure is keeping the string taut so that it doesn't sag and alter the trajectory. Once I had completed the work on half a dozen stains, I was able to identify the point of origin of the blood – the point where all the strings converged. The string reconstruction identified a single site of impact of all the splashes: where the woman's head would have been when she was struck.

It was evident that all the bloodstains I was examining were part of a single pattern and had occurred during the same assault. It was also clear that the only way that all those bloodstains in various positions on the patio door could have got there was if the door had been opened and closed at various times during the assault.

If Scott's scenario was correct and someone had come into the house, carried out the assault and then left, why would they have closed the door halfway through the attack? It made far more sense for someone to have carried out the assault, opened the door in order to make it look as though an attacker had come in, and then struck a further blow or two.

It was also easy to conclude that the reason Scott had not had any blood on her — other than a small amount on her nightdress — was that she had removed her clothes, come downstairs to carry out the assault then headed back upstairs and jumped into the bath to remove the evidence. She had attempted to remove the spots of blood on her nightdress by hand-washing the garment, but this had served only to make her appear more guilty still.

At least partly on the basis of my findings, Jayne Scott was subsequently charged with causing and inflicting grievous bodily harm.

By the time the case came to the Central Criminal Court, a clear motive for the attack had emerged. It transpired that Scott and Fox had been in a sexual relationship with each other for some time, though they had attempted to keep it quiet to avoid potential prejudice at their respective workplaces.

However, although the pair lived together and continued to

share a bed, the passion between them had cooled and they had not had sex for more than a year.

Craker had inherited a house in Barnet a few weeks earlier and had invited Scott and Fox over to help her renovate it. On the first occasion, Scott had been unable to attend, as she was visiting her family. Fox went alone and, over the course of the weekend, she and Craker ended up in bed together.

In the days that followed, Scott found out what had happened. She asked Craker to give up this new relationship, but she refused, telling Scott that she would not be intimidated. Despite the obvious tension the revelations had caused, however, both Fox and Scott agreed to stay at Craker's house the following weekend.

This back story ensured that the case received plenty of publicity, with both the broadsheets and the tabloids running banner headlines about the 'Lesbian Love Triangle' and eagerly recounting every detail they could find out about the love lives of the three women.

When the trial began, it was clear that Scott was sticking to her story of an unknown intruder being responsible. Her defence was bolstered by the discovery of five unidentified fingerprints at the crime scene and the statement of a doctor who testified in her defence that the damage to Craker's brain would have made it impossible for her to remember anything.

Craker herself had now emerged from her coma and attended court to give evidence. Unable to read or write, confined to a wheelchair and able to speak only a few stuttered, staccato words at a time, it was easy for the defence to make the case that her memory of what had taken place that morning was incomplete.

The day before I was due to attend court to give my own testimony in the case, I dropped a glass bottle containing a liquid blood sample and, while clearing up the mess, accidentally cut myself on a small sliver of bloodstained glass.

The standard MPFSL procedure in the event of such a 'needlestick' injury was to send a sample of the blood over to the virology department at nearby St Thomas's Hospital to be tested to ensure it did not contain any infectious diseases. This was duly done.

The following morning, just as I was about to leave for the Old Bailey, I received a call from St Thomas's informing me that the sample had tested positive for Hepatitis B. The hospital informed me that I needed to have an immunoglobulin injection within 24 hours of the original injury to prevent me developing the condition – so there were only a few hours left before it was too late.

I was due in the witness box within the hour and didn't expect to be there long, so I decided to head to court first and stop off at the hospital on my way back. It turned out to be the wrong decision. Although the usher repeatedly assured me that I would be required in court at any moment, the minutes quickly turned to hours and the deadline for my injection started to get uncomfortably close.

I told the usher that if I didn't get called in the next half-hour, I was going to have to leave – an unheard-of insult to a judge. Time passed but, despite further reassurances that I would be in the witness box in the next few minutes, I eventually had to make up my mind. Would I risk upsetting the judge, or risk suffering weeks of nausea, vomiting, fever and jaundice as a result of Hep B? It was a no-brainer. I left court and made my way to the hospital.

Arriving at St Thomas's, I went straight to the virology department and was handed a vial containing 25 millilitres of the immunoglobulin, then told to go down to Accident and Emergency and request one of the nurses there to administer the injection. I did as I was told and was directed to a small curtained-off cubicle and asked to wait. And wait I did. In spite of a few reminders, it was over an hour before a flustered nurse rushed in to see me.

She was feeling flustered not because she felt bad about keeping me waiting but because, as it soon transpired, a very irate Old Bailey judge had just phoned the hospital to ask where his witness was, as the court had been waiting more than an hour.

Instructed to drop my trousers immediately – immuno-globulin is administered into the muscle of the upper thigh – the nurse proceeded to inject all 25ml in a single, rapid movement before announcing that I was discharged and pointing me towards the front entrance, where I could hail a taxi.

Although it may not sound like a great deal, 25ml represents quite a large intra-muscular injection to receive all at once, and just as I passed through the main hospital doors its effect manifested itself in dramatic fashion.

My left leg gave way and I fell, in my smart suit, the whole length of the hospital's concrete steps. I was more embarrassed than injured, but my best suit – the one I always wore for court appearances – had come off less well. I had dirt and grubby stains all down one side and, worst of all, my trousers had split from front to back. I didn't have time to go home to change – the judge was already furious.

The witness box hid much of my 'embarrassment', but there was sufficient damage on show to occasion a raised eyebrow from His Honour and for me to have to apologize for my appearance. He was not amused.

I gave my evidence in chief and waited in the witness box for the cross-examination to begin.

The barrister shuffled through the wad of papers in front of him before looking over at the witness box and fixing me with a hard stare. 'Dr Silverman . . . oh, wait, you're not actually a doctor, are you?'

I'd always known that failing to finish my PhD would be

thrown at me at some point, but on this occasion I didn't mind one bit. I now knew the way the system worked, especially when it came to giving evidence in court.

The defence had already seen my statement and heard my evidence. They knew that my findings were beyond question and fitted in with the case presented by the prosecution far better than any scenario the defence could come up with. But cross-examination isn't always about opposing evidence, it's about finding a way to bring the jury round to an alternative point of view.

We were back on that film set and the barrister was trying his hardest to take the whole production in the direction that he wanted it to go. One thing was certain, however: if his first move was to attack me personally rather than my findings, it meant only one thing – he really didn't have much of a case.

There was more showmanship to follow. At one point, the defence barrister asked whether the presence of a particular bloodstain could give an indication as to whether the attacker was left- or right-handed. I explained that it simply wasn't possible to say, as it would depend on where they were standing when the attack took place.

If the attacker was right-handed, the patterns would have been the result of blows coming from a particular location, but if they were left-handed, identical patterns could have been created, provided they were standing in a different location.

The barrister continued to push me to choose one scenario or the other, but I refused. From my previous experience in the witness box I knew all too well the dangers of making statements which I could not substantiate with science, and so I refused to be taken down that path.

It turned out that a later witness for the defence – who was indeed a medical doctor but had no training in blood pattern analysis – was happy to say with absolute certainty that the blows could only have been struck by a right-handed attacker. The jury was then told that the accused, Jayne Scott, was left-handed. Or visa versa. I can't remember which!

It was another lesson in the ways the courts work and how the jury responds to evidence that is presented to it. My big fear was that I had come across as indecisive because I was erring on the side of caution. The defence witness, on the other hand, had spoken with absolute conviction and seemed certain of what he was saying, and I could understand why the jury might have preferred his evidence to mine.

In the end, it wasn't a problem. Despite the defence putting up a good fight, the case against Scott was a solid one and, after a few hours' deliberation, the jury found her guilty on all counts. She was sentenced to seven years' imprisonment.

Scott's big mistake was attacking Craker a final time, having already tried to dress the scene to make it look as if an intruder were responsible. Such scenarios are not uncommon.

People often underestimate the amount of effort needed to kill someone, or someone's will to live. Murder victims may twitch, making it appear as if the victim is still alive and prompting the killer to launch a fresh attack to finish them off.

I'd known of such cases, including that of the murder of actor Peter Arne. Although not my case, I had assisted one of my colleagues by using my O-level Art to put together some crime scene sketches.

Arne's body was found in the hallway of his flat, surrounded by spots, splashes and smears of blood across the floor, walls, ceiling and doors. More blood had splashed its way into the bathroom.

Considering the nature of the struggle that must have taken place, the room itself was relatively undisturbed. There were two coffee mugs on a trolley, two cigarette ends in an ashtray and a half-empty jar of honey on the floor. Near the door was a rolled-up sleeping bag and a rucksack, both of which were stained with blood. The rucksack was opened and found to contain a shirt and a towel, both of them also bloodstained.

The upper part of Arne's body was lying face downwards, extending from the hall into the small kitchen. Long streaks of blood on the carpet indicated that he had been dragged into that position. Heavy blood staining on the lower parts of the kitchen walls had radiated out from his head. On the kitchen floor were a knife and a wooden stool, both also heavily stained with blood. In the kitchen sink there was diluted blood, and there were more diluted bloodstains on the sink surround and on a dishcloth.

The bloodstains told a clear story (see pages 1–2 of the second picture section for my scene sketches and photographs). Arne was first attacked in his living room. The blow that was struck there was a relatively minor one which caused an injury that bled, but not profusely. Arne had left the room, seemingly followed by his assailant, who had taken a log from the fireplace that he intended to use to continue the attack.

More blows were struck as Arne made his way out of the front door and into the communal hallway. The stool had also been used as a weapon. Present were distinctive patterns of arterial spurting, a sign that Arne's throat had been cut before he was dead.

At the time of his death, Arne was living with 24-year-old Tom Jackson, an aspiring model who had been using the flat as a base while taking his portfolio to various modelling agencies. Arne also had a friend, 44-year-old John Ryan, who lived near by. Both men were found to have rock-solid alibis and were quickly eliminated from the inquiry, though Ryan did give police what would turn out to be a vital new lead.

Ryan related that Arne had recently befriended an Italian man who was down on his luck. Ryan had met Arne one day, and he had explained that he was on his way to Hyde Park to take the man some sandwiches and had invited Ryan to accompany him.

At the park, Ryan had been introduced to the bearded, strongly built man, but not given his name. The man spoke no English and, since Arne could not speak Italian, the two managed to carry on a conversation in French.

The morning of the day he had been murdered, Arne had been at BBC Television Centre in White City to be fitted for a costume for a part in *Dr Who*. He had returned home around midday, where eyewitnesses reported having seen a well-built man with a beard eating from a jar of honey just outside number 54, apparently waiting for Arne.

It also emerged that, during the struggle, the entry phone to Arne's flat had been knocked off its holder, allowing any-one passing the entrance to hear the brutal fracas. It was later discovered that, although several passers-by had heard Arne's desperate last moments, none of them had chosen to alert the authorities, all electing to leave the responsibility to someone else.

The alarm was finally raised by Eva Brava, a nursemaid from the Philippines who worked for a family living in the flat directly above Arne's, who stumbled across a bloodstained log taken from Arne's fireplace, along with his wallet and a number of other items, in the communal hallway.

The killer wore one set of clothes to bludgeon and drag his victim from there to the kitchen in the centre of the flat, then, thinking he had killed Arne, changed into fresh clothes which he had brought for that purpose. As he was dressing, he must have discovered that his victim was still alive, which must have caused him to panic and put his tracksuit bottoms on back to front – leading to a second set of blood distribution marks on his new set of clothes as he launched another attack.

This change of clothes was crucial in helping to show that the killer had planned the murder in advance and that it had therefore been premeditated.

Just three days after the murder, the attacker's body was found floating in a nearby river. His clothes with the second set of bloodstains were found neatly folded on the bank. It was speculated that, it having been a hot day, he had decided to go for a swim in order to wash off any other traces of blood, had been caught by the current and drowned. The mystery of why he decided to kill Arne died with him.

Sometimes it's the absence rather than the presence of blood that's the key to solving a case. A few months after the Arne murder I attended a crime scene in the lounge of a flat in a London tower block.

The victim's body was still *in situ*. He was lying with what was left of his shattered face upwards on a settee, his head on the armrest nearest the door and his body on the seat. The heavy, directional blood staining covering the short wall by the door had clearly emanated from the area of the arm of the settee.

According to the suspect, he and the victim had been standing face to face in the lounge having a furious argument, having been drinking all afternoon. The victim then attacked the suspect who, in self-defence, picked up a large shovel that was (conveniently) standing by the gas fireplace, and struck the victim with it. The victim then fell back on to the settee, where the suspect said he had struck him once, or possibly twice, more.

The problem, however, was that the bloodstain patterns I found at the scene did not quite fit the claimed scenario.

For one thing, it was clear that there had been more than two blows struck with the shovel, which had then been replaced by the fireside – not what you'd expect in a case of self-defence. As the scene did not conclusively support or refute the suspect's self-defence scenario, I turned to the suspect's clothing, which had been submitted to the laboratory.

The suspect had been wearing a silver-grey two-piece suit at the time of the incident. I carefully examined the front and back of both jacket and trousers, and ringed each spot of blood with a yellow wax crayon to make it more easily visible. The resultant blood pattern was intriguing, to say the least.

There was heavy staining of blood splashes down the front of the left leg, on the right sleeve and cuff, on the back of the left sleeve and on the left and right front of the jacket. But, curiously, no blood at all was present on the front of the right leg or the front of the left sleeve. Stranger still, there was a horizontal band free of blood staining across the left front of the jacket, with heavy bloodstains above and below it. The heaviest staining was on the right cuff and the front of the jacket, indicating that these areas were closest to the point of impact (compare the two photos on page 3 of the second picture section).

Looking at the clothing lying there on the laboratory bench, the blood on it seemed inexplicable. However, among the eclectic equipment of the forensic science laboratory was a shop-window dummy, kept for the express purpose of displaying clothing as it is worn. I dressed the dummy in the suit, positioned the arms and feet – and all was revealed!

It was the void areas (the areas in which there was no blood) which were important, in particular the blood-free areas on the front of the right trouser leg and the band across the left jacket front. In a face-to-face confrontation – or, indeed, if the victim had fallen on to the settee and the assailant was facing him (as described by the suspect) – there would have been no rational explanation for the lack of blood splashing in those two areas when all else was heavily stained.

However, if one postulated a scenario in which the victim was already lying with his head on the arm of the settee (sleeping, perhaps) and the assailant had entered the room, carrying the shovel from outside (a far more logical place for a shovel to be located, anyway), and struck the prone victim from behind the settee – then the void areas became clear.

Putting the clothes on the shop-window dummy had made it aparent that the assailant had been standing holding the handle of the shovel in his left hand while wielding the blade with his right. He would have then leaned across the end of the settee to strike the head of the victim, in so doing shielding the right leg of his suit with the settee. In order to lift the weapon high enough over the back of the settee, the assailant's left arm would have fallen naturally across his chest, protecting the left front of his jacket.

The blood staining patterns thus made it clear that this was not a case of self-defence, as had been claimed, but rather a cold-blooded case of premeditated murder.

6

AROUND THE SAME TIME THAT I WAS DEALING WITH THE
'lesbian love triangle' case, Alec Jeffreys, a 34-year-old genetics
researcher at Leicester University, had serendipitously made
a discovery that would for ever change the field of forensic
science and criminal investigation, not just in the UK but
throughout the entire world. Jeffreys had been studying genetic
markers in the DNA of family members in an attempt to see if
he could trace certain inherited illnesses and diseases, such as
cystic fibrosis, through lineages.

DNA was first discovered in 1869, but the fact that the
molecule played a key role in genetic inheritance was not
discovered until 1943. A decade later, molecular biologists
James Watson and Francis Crick, working at the Cavendish
Laboratory, Cambridge University, England, developed a
model for the structure of DNA – a spiral consisting of two
strands wound around each other in the form of a double helix.
For this work, which they published in 1953, and subsequent
work, they were jointly awarded the Nobel Prize in 1962, with
Maurice Wilkins.

Deoxyribonucleic acid (DNA) is a huge molecule containing the genetic instructions that dictate the development and functioning of virtually all living organisms; it is the blueprint for who and what we are. The DNA sections which carry this genetic information are called genes, and are 'super-coiled' into structures called chromosomes. The DNA itself is composed of four building-block molecules ('bases') – adenine, thymine, guanine and cytosine – which are usually referred to by their initial letters, form exclusive pairs (A with T; G with C) and repeat millions of times throughout each DNA strand, making different patterns. Some of these sequences of bases at a particular site (locus) on the DNA molecule code for the genes; other sequences have structural purposes, or are involved as switches regulating the use of genetic information. Some genes have multiple forms, known as alleles, at a single site. However,

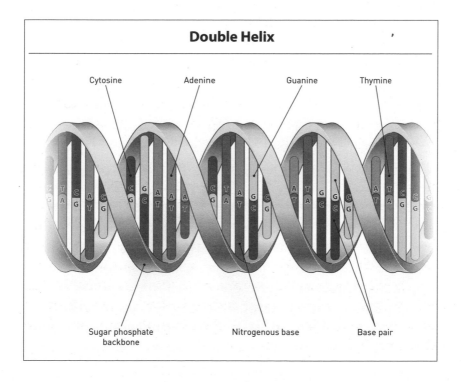

Double Helix

Cytosine   Adenine   Guanine   Thymine

Sugar phosphate backbone   Nitrogenous base   Base pair

over 90 per cent of DNA (thought in 1984 to have no genetic purpose) is between the genes and contains highly variable sequences of base codes.

One major source of variation within DNA is a particular type of repetitive sequence (or 'stutter') known as a Variable Number Tandem Repeat (VNTR). VNTRs are fragments of DNA that are composed of identical units of base pairs (between 20 and 100 base pairs in length), which are repeated anything up to 100 times. The overall molecular weight of a VNTR is determined by the number of times the units are repeated. This highly variable region of the DNA molecule is known as a 'minisatellite'.

In 1984 Jeffreys and his colleagues were concentrating their attention on these minisatellite sections of human DNA, as they contained more apparent dramatic variations than other strands (known as core sequences) and would therefore make better markers for tracking the positions of genes.

A particular genetic stutter could be unique to an individual, Jeffreys realized, and so he devised an experiment to see if he could count the repeats in different individuals and their relatives, as well as in animals such as seals, mice and monkeys.

Using various samples of genetic material, including some blood from his technician and her parents, he first set about breaking open the cells in order to extract the DNA inside. (In the case of blood samples, only white blood cells can be used for DNA analysis – red blood cells have no nucleus and therefore contain no DNA. See page 5 of the first picture section.)

The DNA strands were then 'cut' at specific points, using reagents called restriction enzymes. Restriction enzymes were specifically chosen to correspond with sites on either side of the minisatellite positions on the DNA. In this way, the minisatellites were excised and the inherent length variation of their VNTRs could be assessed. Fragments of varying length were placed on a gel matrix and had an electric current passed through them in order to distinguish the longer from the shorter ones. (As mentioned previously, this process is known as electrophoresis.) The latter move more quickly than the former towards the positive electrode.

The DNA fragments thus separated were then transferred to a nitrocellulose or nylon membrane using a technique known as Southern Blotting, during which they were 'fixed' in place with ultraviolet light. A collection of different synthetic DNA probes, each radioactively labelled, were attached to the DNA fragments and high-speed X-ray film placed in contact with the membrane and exposed. On development of the film, a dark band was formed at any point where a radioactive probe had been attached, producing a photograph showing dozens of bands.

As the technique was designed to examine variation (polymorphism) in lengths of DNA formed using restriction enzymes, it was termed Restriction Fraction Length Polymorphism (RFLP) analysis. And as a collection of radioactive probes was used to bind to VNTRs at multiple locations on the DNA, this process was termed multilocus probing (MLP) – soon to be dubbed, confusingly, 'DNA fingerprinting'.

On Friday 7 September Jeffreys placed the results of his latest experiment into a photographic developing tank and left it over the weekend. When he entered his laboratory the following Monday morning and removed the X-ray film from its tank, he found a complex array of blobs and lines. His first reaction was that the whole thing was nothing but a mess, but as he looked at the plate more closely, he began to see something more.

The film showed a sequence of bands, each representing different numbers of DNA repeats in the various individuals and animals in the experiment. These appeared as a sort of bar code, similar to those seen on supermarket products.

Crucially, every individual in the sample had a different bar code and could be identified with precision. Jeffreys could even establish relationships between family members: the bands of the DNA supplied by one of his technicians were a composite of her mother's and father's, for example. Even the non-human samples indicated that individual animals could be identified in this way.

It appeared that every organism and person in the experimental sample had a unique pattern of these DNA fragments in their minisatellites (the only exception later identified being identical twins). Jeffreys had developed a DNA-based method not only for biological identification, but also for discerning family relationships. He rushed out of the X-ray developing room to tell his team that they were on to something really exciting.

'It was an absolute eureka moment,' he said later. 'It was a blinding flash. In five golden minutes, my research career went whizzing off in a completely new direction. The last thing that had been on my mind was anything to do with identification or paternity suits. However, I would have been a complete idiot not to spot the applications.'

Jeffreys gathered his staff and they all began brainstorming, trying to think up practical uses for the new technology they had stumbled across. The fact that the DNA analyses of offspring were a composite of those of their parents ensured that paternity cases were high on the list of options, but the team also wondered if the technology could be used to identify criminals, using the blood and other body fluids so often left behind at crime scenes.

Jeffreys understood that he could not pursue the idea any further until he could be certain whether DNA was robust enough to survive outside the human body. No one had ever asked the question before, so it was a complete unknown. For all he knew, the chromosomes could fall apart once the cells died, or could be destroyed when a fresh bloodstain had dried up. There was only one way to find out.

He spent the next two days cutting himself and leaving blood marks around the laboratory. Then he tested those bloodstains and found that their DNA was intact.

A paper about DNA fingerprinting was written by Jeffreys and his team and was published in the scientific journal *Nature* in March 1985. News of the discovery was subsequently reported in several newspapers.

The following month Jeffreys received a letter from Sheona York, a London lawyer. She had read the coverage of DNA fingerprinting and wondered whether this technique could help sort out a tricky immigration dispute involving a family from Ghana.

The youngest boy had gone back there for a holiday and returned with a United Kingdom passport which appeared to have been tampered with. Immigration authorities suspected that the boy was a substitute, perhaps a cousin, trying to sneak into the country using a relative's passport.

The family had been subjected to traditional blood grouping, and the results of tests for ABO, Rhesus and 16 other proteins and enzymes showed without doubt that the woman and the boy were almost certainly related. Due to the limitations of blood grouping, however, the test could not say with any certainty whether the woman was the boy's mother or simply his aunt.

Although there were some additional complications – the boy's father was no longer in the UK, so he was not available for testing – York wondered if the new procedure could prove that the boy was indeed the mother's son.

At first Jeffreys was unwilling to get involved, comparing

the case to a jigsaw puzzle with too many missing pieces. However, after discussing the case with his wife, he decided to make an attempt. The test proved far more conclusive than he could have ever hoped. First he managed to reconstruct the DNA fingerprint of the missing father by using DNA from the woman's three other children. He was then able to show that every genetic characteristic of the disputed boy matched the mother or father. As a result, the immigration tribunal, once the technology had been explained to them, dropped the case and allowed the boy back into the United Kingdom as a full citizen. The heart-warming case also resulted in another round of publicity for Jeffreys and his new technology.

Soon, so many immigration and paternity cases were coming in to the university that a separate company had to be set up to deal with them. In 95 per cent of cases of immigrants who had been refused entry to the UK, the results showed that they were indeed blood relatives of existing UK citizens and were therefore entitled to citizenship.

By now, the Home Office was taking a greater interest. Two scientists from the Central Research Establishment at Aldermaston (part of the Home Office Forensic Science Service) were despatched to Leicester to assist Jeffreys in determining whether the technique could be applied to crime scenes.

In December 1985 Jeffreys, along with Peter Gill and Dave Werrett of the HOFSS, wrote to *Nature* and noted that they had been able to obtain viable DNA samples from four-year-old blood and semen stains on cotton cloth. They had also developed a technique for separating sperm nuclei from vaginal cells on semen-coated vaginal swabs to enable positive identification of a male suspect. The letter noted: 'It is envisaged that DNA fingerprinting will revolutionize forensic biology, particularly with regard to the identification of rape suspects.'

★

The following year a call came from Leicestershire police, who were investigating the rape and murder of two schoolgirls who lived in the village of Narborough, just outside Leicester.

The body of Lynda Mann, 15, had been found in the grounds of Carlton Hayes psychiatric hospital in November 1983. Forensic examination of semen samples showed that it was a type found in only 10 per cent of men, and was from someone with type A blood. However, the police did not find a suspect for comparison, so the case quickly came to a dead end.

Three years later, another 15-year-old, Dawn Ashworth, was similarly sexually assaulted and strangled in the nearby village of Enderby, and semen samples showed that the attacker had the same blood type as the first.

Richard Buckland, a local 17-year-old with learning disabilities who worked at Carlton Hayes psychiatric hospital, had been spotted near Dawn Ashworth's murder scene and was brought in for questioning. He seemed to know several unreleased details about the body and subsequently confessed to murdering her. However, he insisted that he had not killed Lynda Mann.

The police asked Jeffreys to use his new technology to determine whether Buckland was responsible for both murders.

The X-ray film that he produced was covered in a pattern of black bands which clearly indicated that the semen from both girls came from the same man, as expected – but that Buckland's DNA was completely different. Despite the police being convinced they had the right suspect, the tests showed that Buckland had to be innocent.

Jeffreys initially thought that something in the DNA profile was flawed. To find out whether this was the case, he turned to the two Home Office scientists, Gill and Werrett, who had assisted with his early research and asked them to repeat the tests independently. They obtained the exact same result. When

Jeffreys passed the information on to the police investigators, they were initially very disappointed but in the end accepted the validity of the results. Buckland was set free. The very first criminal investigation to involve DNA evidence had resulted in an exoneration, rather than a conviction.

Like many disturbed and emotionally vulnerable suspects before him, Buckland had made his false confession simply because he had been unable to handle the pressure of an interrogation. 'I have no doubt whatsoever that he would have been found guilty had it not been for DNA evidence,' Jeffreys said later. 'That was a remarkable occurrence.'

In January 1987, with the real killer still on the loose, the police decided to place all their trust in the new technology and announced the launch of the world's first DNA-based manhunt, asking for blood samples from all men living in the three villages surrounding the locations where the bodies had been found. In all, tests on nearly 5,000 men between the ages of 18 and 35 would be required.

There was just one small problem. The moment they saw the commercial potential in Jeffreys' discovery, the Lister Institute of Preventative Medicine, of which Jeffreys was a Fellow, applied for a patent and entered into an agreement with leading UK chemical company ICI. They persuaded Gill to sign away any intellectual property rights he might have had. A brand-new subsidiary of the institute, Cellmark Diagnostics, had been formed with a view to exploiting DNA fingerprinting to its fullest extent. Tests were being offered at around £110 each – four times the cost of blood grouping – so the mass screening represented a lucrative opportunity for the fledgling company. Panic ensued at the Home Office, with senior mandarins asking how this had happened.

However, having assisted Jeffreys in confirming the results of the Buckland test, Gill and Werrett were confident they had all the information and skills they needed in order to carry out the DNA analyses themselves rather than paying ICI to do so. For a time, it looked as though the dispute might end up in court. ICI was keen to protect its interests but also aware that taking action would open the company to criticism that it was trying to profit out of a murder investigation.

Ultimately, the Home Office claimed 'Crown Privilege' (a legal manoeuvre since renamed Public Interest Immunity), which allows the government to acquire a technology regardless of intellectual property rights if it is felt that it is in the public interest to do so.

The mass test took more than a year. First, blood grouping analyses were carried out to eliminate those who did not match the traditional blood groups in the samples found in both victims. These tests were not only relatively cheap, but results could be obtained almost overnight. The 10 per cent of men who remained of interest after this initial procedure were subjected to a full DNA test, something which took weeks and sometimes months to complete.

Most men gave blood willingly – the response rate was around 98 per cent – but others needed a little more persuasion. One man, on the run from an assault charge, called the police team in charge of the operation and agreed to provide a sample only if the detective inspector gave his word that he would not be arrested for the outstanding warrant.

With an agreement in place, the man attended his local police station and, in the presence of the DI, gave the sample. As he made to leave, the DI told him that he couldn't just walk away after all. The man was furious and prepared to fight his way out. The DI then calmly explained that he couldn't possibly let the man go without buying him a drink at the local pub first.

One man in particular went to even more extreme lengths to avoid giving a blood sample. Bakery worker Colin Pitchfork told several friends that he was scared of both the police and needles and did not want to take part in the tests. He offered money to anyone willing to give a blood sample on his behalf and, finally, his friend Ian Kelly agreed. For a payment of £200, Kelly took a modified version of Pitchfork's passport and gave the required blood sample, which led to his friend's name being eliminated from the list of suspects.

A few weeks later, Kelly couldn't help bragging about what he'd done while drinking with a group of friends in the pub. One of the friends was deeply disturbed by what she heard and grew even more anxious when, a week or so later, news of the first court case to involve DNA evidence was featured in the newspapers. The case involved a man accused of unlawful intercourse with a 14-year-old mentally handicapped girl who had given birth to his baby. Jeffreys himself was quoted as saying, 'The use of the test in a court case is exciting for us. It is an historic occasion.'

The woman who had overheard Kelly's confession had a friend whose son was a PC, but when she tried to call him she found he was on holiday. It would be six weeks before she would finally pass on the information.

Kelly was arrested and quickly broke down, admitting what he had done. He explained that Pitchfork had cut out the photograph in his passport and replaced it with one of Kelly, had driven his friend to the school where the samples were being taken and waited outside the gates until he came out.

Police checked the signature 'Pitchfork' had produced while giving his blood sample with one from the real Pitchfork's housing records. They did not match. He was subsequently arrested and his DNA was found to be a perfect match for both murder victims. Kelly was charged with conspiracy to pervert the course of justice, while Pitchfork was taken into custody.

Pitchfork knew enough about DNA to know the evidence against him was irrefutable – which was why he had gone to such great lengths to avoid taking the test in the first place. As he sat in the interview suite, there was no longer anything for him to hide.

'Why Dawn Ashworth?' asked one detective.

Pitchfork simply shrugged. 'She was there and I was there. Opportunity.'

A long-time flasher, Pitchfork later confessed that on the night he murdered Lynda Mann he had dropped his wife at college and gone out looking for women to expose himself to, with his young son in his car seat as he drove from Leicester to Narborough.

He spotted Lynda, parked his car then waited beneath a street lamp to expose himself. Most of the time when he did this, women simply ignored him or ran away. But Lynda spotted Pitchfork's wedding ring and said, 'What about your wife?' She then ran off down a local footpath called the Black Pad.

After raping her, Pitchfork realized that he was wearing an earring, which could help to identify him. He also said that, at that time, he was planning to move to Littlethorpe, near Narborough, and thought that Lynda would almost certainly see him in the area. To prevent her from identifying him, Colin Pitchfork strangled her with her own scarf.

Three years later he was riding a motorbike, cruising for women to flash, when he spotted Dawn Ashworth. He raped her in a field next to a footpath and, once again, strangled her in order to prevent her from identifying him.

He was sentenced to life imprisonment.

All the staff at the Metropolitan Police Forensic Science Laboratory – and, in particular, those in the biology department – followed the news of the discovery of DNA fingerprinting and its first use in a criminal trial with great interest. We all realized there was potential there, but few people imagined it would become the all-encompassing game-changer it ultimately turned out to be.

Although DNA was better at identifying individuals, the technique in those early days was extremely slow and required a lot of genetic material in order to work. By this time, blood grouping had become increasingly sophisticated, with around 32 separate groups identified, increasing the chances of being able to distinguish between the body fluids of two individuals.

In my mind, DNA seemed to be just another form of grouping, albeit a more accurate one. Personally, I looked forward to getting to grips with it, but didn't think it was going to change very much in my day-to-day life at work. I couldn't have been more wrong.

7

SERIAL CRIMES, MOST NOTABLY MURDER, RAPE AND ARSON, are all staples of TV police dramas and detective novels but, in reality, they are few and far between. Even in America, where the problem is more endemic, serial killings account for less than 2 per cent of total homicides.

Serial rapes and arson attacks are far more common but, still, for the average real-life investigator, the opportunity to investigate such a crime occurs at most just once or twice during the course of a typical career. One of the greatest challenges in the investigation of these types of crime is deciding whether or not a new offence is part of the 'series'. Incorrectly including a copy-cat offence or, indeed, an unconnected incident, can distort the entire series and subsequent investigation and lead to totally inappropriate paths of inquiry, dead-ends and wasted time and resources.

Identifying a series attacker is usually a case of incidents having been committed by similar patterns of method – modus operandi or MO – as, generally, that is all there is to go on. When a series of rapes took place in south London in the mid-1980s, the MO of all the attacks and description of the man alleged to be responsible were similar.

Although DNA was still in its infancy, another part of the MO was that the man concerned was very forensically aware. He would never leave anything behind at the scene, usually choosing to ejaculate on to an item of clothing which he would then take away with him.

Everald Irons had first appeared on the police radar for these crimes after a woman claimed she was raped in Streatham in December 1985 by a man fitting his description. Three months later, a young prostitute from Leeds was waiting for a girl-friend in Streatham High Road when she was grabbed by a man with a knife. However, a routine police patrol saw and, assuming the woman was about to be mugged, stepped in and arrested the man. It was Irons.

Detectives spotted the similarities between the two attacks, and the first rape victim picked out Irons from an identity parade which, unusually, took place on an escalator at Kennington underground station. He was charged with rape, but the case collapsed at the ensuing committal at Camberwell Magistrates' Court on a technicality. Irons had in fact only attempted to rape her and had forced her to perform an oral sex act, but the woman had claimed he had forced himself on her. She later broke down in the witness box and admitted she had lied under oath because she wanted him convicted of rape. Irons pleaded guilty to common assault and was given an absolute discharge.

In the ensuing months, there were at least 18 rapes and indecent assaults in south-west London. All but one of the victims were gagged and had their hands tied behind their backs during late-night attacks after being stalked from behind and taken to wasteland or dark alleyways to be assaulted. The victims described their attacker as tall, black and bespectacled.

They all recalled that, after raping them, he had untied their hands before making his escape.

One victim begged not to be tied up, but the man told her: 'I won't hurt, I swear on the Holy Bible. But I've got to do this. I'm into bondage. It turns me on.'

Irons, an only child, had been brought up by his mother in a strict Baptist household. (In deference to his upbringing, he never carried out any attacks on the Sabbath.) A bondage magazine was found under his bed during one of his early arrests, but it was returned to him when he was freed without charge.

An artist's impression, which bore a striking resemblance to Irons, was drawn up and police received hundreds of calls in response when it was made public. Several of those callers named Irons, and he was arrested, but none of the alleged victims could pick him out of the identity parade, so he was released 36 hours later.

In March 1987 a woman went into Wandsworth police station to report a car accident and saw the artist's impression that had been drawn up months earlier. She told officers that the man in the poster had raped her the previous year, but she had not reported the crime at the time. Irons was arrested once more, but yet again the victim failed to identify him and he was allowed to go free.

Police were determined to put a stop to the series rapes, and a new detective superintendent was brought in to review the case. It soon became clear that, despite the failures to fully identify him, Irons was still very much the prime suspect. A huge surveillance operation using a team of 12 officers, some of them posing as Irish labourers, others strategically stationed in a variety of vehicles, was set up to monitor his every move from 21 April.

During the course of the next 26 days, Irons was seen on

several occasions driving his wife's car on 'circuitous routes' in the Tooting and Balham areas of south-west London, taking him past or near the scenes of several attacks. On one particular day he was seen by police to stop near Tooting Bec Common and to disappear into some bushes with a belt over his shoulder. He emerged shortly afterwards, but there was no evidence of any crime having been committed so no action was taken.

As the costs of the operation began to mount up, the surveillance was withdrawn. Just a few days later, on 27 May, Irons carried out his next attack. His victim was returning home when she saw a man appear out of the shadows near Tooting underground station.

He told her: 'Don't scream, all I want is your money.' He pushed her into nearby marshland, gagged and raped her, then forced her to perform oral sex on him.

The timing of the attack seemed to confirm what many on the police team already suspected: that Irons was getting inside information about the case.

Irons had met his wife, Anne, at a Baptist church in the late 1970s. The pair had started going out a few years later and were soon engaged and then married. By that time, Anne had joined the Metropolitan Police, working in the youth and community section. A born-again Christian, she admitted that on many occasions before and after their marriage, Irons would leave her alone at their flat and go for a drive on the pretext that he needed 'space'.

Anne Irons had provided her husband with several alibis, stating he was with her when some of the rapes took place. She also claimed that a number of times and dates could be confirmed by entries in her police diary, which had previously been submitted to the Metropolitan Police Laboratory to make sure it had not been altered but had been returned to her when the charges against her husband were dropped.

The rape on 27 May was reported immediately and Irons was arrested the next day. This time, his victim picked him out of the line-up and, at long last, a case against him could be pursued. Once again, Anne Irons provided her husband with an alibi but, this time, when her police diary was submitted, it was found to contain a number of false entries.

The analysis of handwriting is one of the oldest forms of forensic science around. Cyril Cuthbert, the sergeant who helped found the Metropolitan Police Laboratory, made his name by managing to show that a particular document had been altered, and many techniques exist to establish whether such tampering has taken place.

In cases where forgery is suspected, a suspect will usually be asked to give writing samples. Although it is possible to alter one's handwriting deliberately, doing so takes a conscious effort, while writing itself is a mostly subconscious activity.

To prevent cheating, suspects have long passages dictated to them. Listening and altering your handwriting at the same time is not possible for anything more than a brief period, so most suspects end up slipping into their natural handwriting at some point.

It is also possible to reveal otherwise invisible pen impressions by using laser light at different frequencies. In April 2012 police in Devon reported that their forensic laboratory had helped a blind novelist recover 26 pages of a manuscript she had written without realizing that her pen had run out of ink!

Showing that writing has been added to pages at a later date is usually the work of the ESDA – Electro-Static Deposition Apparatus – a vital piece of equipment in the documents section of any forensic laboratory. The machine works due to the fact that the pressure exerted on paper by the tip of a pen creates minor changes in the paper fibres, not just on the page being written on but also to a depth of up to five pages.

The document under investigation is first placed in a high-humidity environment to increase its water content, then positioned on the bronze plate on top of the ESDA and held in place by a vacuum. A film of plastic is put over the document to protect it and tiny glass beads covered in photocopy toner poured on top. A current is passed through and the toner beads, attracted by the disturbances in the fibres, show what indented writing is present.

By comparing the markings on one page with another, it is possible to spot if text has been added after an entry has originally been made.

However, on this particular occasion, none of the sophisticated equipment used by the documents section was required. When Anne Irons' diary was submitted that first time, photocopies had been made of all the pages. When these copies were compared to the same diary the second time, it became evident that several entries had been added, all with the goal of providing alibis for dates of earlier attacks which had emerged since Irons had last been arrested.

Different clothes shed their fibres to different degrees. What someone is wearing can have a dramatic effect on whether textile fibres are of any relevance in their particular case. If someone is wearing nylon, there is little point in looking to see if fibres have been shed. If they are wearing a loose-woven wool such as angora, which sheds a lot and leaves fibres on virtually everything it comes into contact with, the prospects are good.

Fibres transfer, but are lost from the recipient surface over time. If a long period has elapsed before an arrest is made, there may be little of this kind of evidence left. Everald Irons, however, was arrested just a few hours after his latest rape, so we were in luck.

The clothes of both suspects and victims in rape cases are seized, securely packaged by the SOCO and 'taped' at the laboratory. This involves literally using a piece of clear adhesive tape to remove extraneous textile fibres from the surface of each item. Such tapings are carefully labelled with a note of where they were taken – right lower trouser leg, left inside pocket, and so on. To prevent cross-contamination, fibres and tapings from the suspect and victim are never examined at the same workbench and the technician changes his lab coat to work on each one.

Once I know what the victim and the suspect were thought to be wearing at the time of the attack, I can decide which fibres are worth looking for. Firstly, I might test the victim's garments to see how well their fibres shed. This is easily achieved by seeing how readily the garment's own fibres are removed on a small piece of clear adhesive tape. I then remove a few small threads of each of the colours of a garment and mount these as a 'control sample' on a microscope slide in a transparent medium called Xam. This provides the basis for comparison with any fibres found on a suspect's garment.

In the Everald Irons case, the victim of the latest attack had been wearing a woolly dark-coloured top that shed very well. I knew that, thanks to the short amount of time that had passed since the assault, there was a good chance Irons would still have the fibres on his clothing.

Wearing a clean lab coat in a different, designated part of the laboratory, I spread the suspect's jacket out on a large sheet of brown paper. Using six-inch lengths of tape, I started 'taping' the garment one section at a time to remove all extraneous fibres from the surface. As the tapes filled I stuck them down on a sheet of polycarbonate and labelled each one. Once all extraneous fibres had been taken from the whole garment I was ready to search for fibres that might have come from the victim. And by 'fibre' I do not mean a 'thread' but the smallest sub-division it is possible to make of a thread, often so small as to be barely visible to the naked eye.

Having made a mental picture of what the 'control sample' from the victim's garment looked like under low-power magnification, I placed the first of the tapings from the suspect's garment under the microscope. Inch by painful inch I examined the whole length of the taping, removing with a fine pair of forceps and xylene solvent every tiny fibre from the tape adhesive that appeared to be similar to those of the suspect. Each fibre was washed in solvent and mounted individually under glass on a microscope slide. Only then was I ready to make use of the high-power comparison microscope.

This piece of equipment is essentially two microscopes with one eyepiece between them, which allows the user to obtain a split-screen view of the fibres on the left and right stages simultaneously. I mounted the control sample on one stage and the first of the fibres removed from Irons' jacket on the other. This kit takes a bit of setting up to get all the light condensers and optics balanced correctly, but once I'd done this I was able to see, under much higher power magnification, that the fibres on the two stages were completely different. So, on to the next of my great pile of unknown extraneous fibres.

In addition to finding the fibres themselves, it was also important to ensure that I found enough of them to make the case for primary contact. If it was just one or two, it would be too easy for the defence to claim they had been transferred by secondary contact – via a third party (policeman, SOCO or scientist). A large number of fibres from the clothes of the victim on the suspect's clothes, however, would help establish his primary contact and hence possible guilt.

It is sometimes possible to spend weeks on end in this darkened, windowless cupboard searching for the elusive fibres (which may not even exist) with the comparison microscope, going from one to the next and the next until, finally, hopefully, bingo!

The fibres department at the laboratory had recently had success with an armed robber who wore a duck-egg-blue Lacoste jumper that was only available in one of four outlets in London. I was hoping I would be able to find something equally rare to help connect Everald Irons to his most recent victim.

I had enjoyed working in the fibres department during my early training at the laboratory and, as it was one part of my specialist area, I'd had plenty of opportunity to work with textile fibres on a regular basis.

Most frequently, fibres were useful in order to demonstrate primary contact between two individuals, or between the individual and the crime scene, as proposed by Locard's Principle. Sometimes, fibres came into cases because they were found on a knife blade or gun barrel, and it was important to show that they could be matched back to a particular victim's outfit. On other occasions, hair or fibres, or a mixture of the two, were recovered after armed robberies, particularly if the offenders have been wearing balaclavas or stockings on their heads.

Most of the time it was necessary to do a lot of scientific comparisons in order to confirm a fibre match. After low-power microscopic comparison it was necessary to move on to high-power comparisons, which also aid in the identification of the visual colour match and fibre composition (wool, cotton, synthetic, man-made) as well as cross-sectional shape (flattened, round, tri-lobal, multi-lobed, etc). If the fibre under examination matched the control in colour, shape and composition it was time to move on to instrumentation.

Forensic scientists have long been experts at taking existing techniques and adapting them to tiny amounts of material in order to use them in casework. Fibre analysis is no exception. Two instruments – the visible spectrum microspectrophotometer and the infrared microspectrophotometer – had both been altered to produce spectrographs of individual single fibres, without destroying the fibre. Analysis by these instruments can therefore give a detailed colour-spectrum analysis and composition breakdown of an individual fibre, making the chances of a false match extremely slim.

All yarn which is made into clothing is subject to something called dye-batch variation. It is possible to have two items of identical clothing made by the same manufacturer that look identical in colour but are actually composed from different dyes. This can happen for a number of reasons: a change in the dye supplier, or a change in the formula, a new batch from a supplier, a new dye mixer. The final result will be the same colour visibly, but the proportions of individual dyes may be radically different. Most colour mixing is done by hand by a specialist operator who has a highly skilled eye for colour and knows how to match one batch to the next.

Almost all coloured yarn is made up of several different colours, and it's possible to extract the dyes from a single fibre and separate them to see these individual shades. As no test tubes are made that are small enough, we made our own by melting the ends of tiny capillary tubes.

The minute fibres are placed into these using very fine forceps and a microscope. A minuscule volume of solvent is added using micropipettes, again made specially. A range of solvents can be used to ensure that the best dye extraction is achieved. Once extracted, the dye is applied as a single spot to a small, thin-layer chromatography (TLC) plate consisting of a film of a special clay called kieselgur on an aluminium backing plate.

This TLC plate is then placed in a shallow dish of another solvent, which rises by capillary action up through the kieselgur and separates the dye into its component dyes and colours. The colours and positions (Rf values) of the spots on the resultant multicoloured chromatogram are indicative of the parent complex dye.

Cloth made from yarn that is dyed as part of the same batch will have an identical mix of dyes and therefore produce an identical chromatogram. If the yarn was dyed as part of a different batch, the range of colours and Rf values of the component dyes will be dissimilar, even though the original yarn will look exactly the same colour to the naked eye.

I had found fibres on the clothing of Everald Irons which matched, in every respect and by every analysis, with those which composed the victim's cardigan. Normally, it was enough to demonstrate that identical fibres were present in sufficient quantity to ensure that secondary contact was no defence.

Once I had extracted, separated and characterized the dyes from the fibres, the next step was to identify the size of the fibre population in the UK – in other words, to find out how many garments composed of those particular fibres there were in the country. In order to do that I would need to accomplish two things. The first was to track the garment back to the manufacturers to find out how many had been made. The second was to find out exactly where the yarn used to make the garments came from.

The first part of the task was made easier by the fact that we knew that the victim's cardigan had been sold by Top Shop. At the time, the Metropolitan Police Laboratory had a specialist researcher, Ken Wiggins, who had a real gift for this kind of investigation. He got in touch with Top Shop and in a matter of days came up with some extremely pleasing statistics. Although around 5,000 of the cardigans had been delivered to the company, more than 4,000 of them were still in the main warehouse.

Of the thousand that had been distributed, Ken was able to track their location and found that the vast majority of them were still in the retail stores, yet to be sold. By a sheer stroke of good fortune, the victim in the case had been one of the first people in the country to buy that particular style of cardigan. Had the attack taken place a year later, there could easily have been thousands of such cardigans in circulation but, as it happened, there were, at most, a few hundred. It was a very good start.

The next step was to find out how the cardigans were manufactured. Each new batch would have been made from the same yarn and would therefore have the same dye. If we could identify the size of each batch, we would be able further to reduce the likelihood of someone else wearing a similar cardigan and coming into contact with Everald Irons.

Ken traced the cardigans back to a company based in Dublin which was responsible for making the cardigans, and arrangements were made for someone from the police team to go over and interview the factory owner.

It was a specialist area, and some of the questions would need to be extremely technical. There was little point in a police officer going over to interview the owner on his own, and writing out a list of questions wasn't practical, as so much would depend on the factory owner's early responses. To this end, it was decided that it would be best all round if I went over to Ireland as well.

Myself and the detective sergeant from Operation Optic (as the investigation was known) headed out to Dublin a few days later and, after being made to feel extremely welcome by the Garda, made our way over to the garment factory to interview the owner.

Sitting alongside a police officer and gathering information during an interview was a new experience for me. One of the biggest inaccuracies in the way forensic scientists are portrayed on television is that they are routinely shown in the interrogation room, interviewing suspects. In real life in the UK, this never happens, but, even though I was interviewing witnesses rather than suspects, I couldn't help thinking that I was far closer to that dramatized version of the role than most forensic scientists ever got to be.

It turned out that, although the Dublin factory had put the cardigans together, they did so from yarn obtained elsewhere. Many modern garments are made up with yarn from a number of different suppliers, as this allows manufacturers to take advantage of the best possible prices at any one time. In the case of the cardigan in question, we had another stroke of luck. The garment was made with yarn supplied by just one company, which was based in Portugal.

Having taken down the necessary details, the sergeant and I spent the rest of the evening in the company of the Garda, who insisted on schooling us in the delights of two local beverages – Guinness and whiskey.

I was looking somewhat worse for wear when we returned – so much so that my girlfriend commented that, with my skin being deathly pale and my dark suit, I looked as though I had turned into a pint of Guinness myself!

A few weeks later, the case officer and I flew out to Lisbon. The detective sergeant had no jurisdiction outside of the UK so, before we could gather any information, we first needed to visit a local judge to obtain a *commission rogatoire internationale* – a document that would grant us permission to investigate or search for evidence for our case.

There was no reason to think permission would not be granted – such cooperation is essential in a borderless Europe

– but there was always the chance that some paperwork had been filled in incorrectly or that some or other term might have been mistranslated, so it was something of a relief when *rogatoire* was granted.

The day after, documents in hand and in the company of a local municipal police officer who was to act as our translator, we went to see the firm responsible for dyeing the yarn. It wasn't a huge operation – you could barely describe it as a factory. From our point of view, this was a good thing, as it meant we were going to be dealing with smaller quantities of yarn and this would make our job far easier.

The owner showed us around the operation and demonstrated how they spun the yarn and which vats were used to supply the Irish factory. The vats used to dye the batches of yarn were nowhere near as big as they could have been (see photo on page 5 of the second picture section). At four feet high and six feet wide, they held only a few hundred gallons of dye each time. It meant that dozens of individual batches would have to be dyed to produce the quantity of yarn needed to fulfil the Irish orders.

We also met and interviewed the man in charge of mixing the colours, which was done by hand using a series of controls on the front of each vat. Years of experience meant that this man was able to produce yarn that looked identical each time it emerged from the vat, even though different combinations of colours were required to account for differences in environmental conditions and the quality of the yarn being used. We knew that those tiny colour variations made each mixture unique.

What this meant, ultimately, was that even though a large number of the cardigans had been made, at the time of the attack they would have been a rarity on the street. It wasn't possible to work it out statistically – you would have to build in a wide range of factors, such as how many people own cardigans in the first place, what the weather was like on the day, how people's dress habits are affected by changes in fashion, and so on.

What I was able to say with some certainty was that any garments that were made from yarn that was not in that particular batch on the day that particular yarn was dyed would have different characteristics, and that the overall number of cardigans produced from the yarn dyed in that individual batch would be very small. The chances of Irons coming into contact with someone other than the victim who had been wearing that particular cardigan dyed in Portugal on that particular day were tiny.

A few weeks after returning from Portugal, I stepped into the witness box of the Number One court of the Old Bailey to give my evidence.

Although DNA fingerprinting on semen found on the victim had been unsuccessful, I did have traditional blood grouping results giving a frequency of one in 300, and my confidence in the evidence I had gathered that the fibres I had found on Everald Irons were a match for the cardigan of the victim was the highest I had ever known in any fibres case I'd been involved in. I felt certain that it, along with material put together by the other parts of the team, would finally see Irons convicted of his crimes.

With so much at stake, we all expected the defence to put up one hell of a fight, so I braced myself for a long and difficult cross-examination. They did spend a great deal of time questioning my findings, but my confidence that they had no defence grew the moment the defence barrister began to speak.

'So, Dr Silverman . . . oh, wait, you're not actually a doctor, are you?'

Irons was eventually found guilty of six rapes and one attempted rape. He was sentenced to 18 years.

A week later, the police team that had run Operation Optic held a party to celebrate the conviction, inviting me and Ken from the laboratory along as guests of honour.

It had been a difficult and challenging case, but as a result of diligence and teamwork, we ended up with a good result. I enjoyed the feeling of being a key part of the investigation team, an essential link in the chain leading to a successful conclusion.

Although DNA had failed to produce a result in the Irons case, techniques were improving all the time, and it was clear that in the very near future linking suspects to crime scenes would become faster, easier and far more accurate.

The future seemed bright. No one could have possibly foreseen the difficulties that lay ahead.

8

_____

ATTEMPTING TO SOLVE CRIMES WITH BLOOD GROUPING WAS always something of a gamble, as the technique often failed to provide a result. The relatively large amount of body fluids required to carry out ABO tests meant there was never any point in attempting to test tiny spots of blood or semen, because you'd never be able to get a result. The phrase 'insufficient material for grouping' would be the conclusion of dozens of my reports over the years.

When a DNA unit was first added to the Metropolitan Police Laboratory there was a degree of uncertainty about how useful it would prove to be in the long term because, initially, the amount of genetic material required for DNA testing was even greater than that required for grouping. Small amounts of blood were particularly useless when it came to early DNA 'fingerprinting' technology. As I mentioned earlier, red blood cells do not have a nucleus and therefore have no DNA. White blood cells do have DNA, but while red blood cells make up between 40 and 50 per cent of the volume of a typical drop of blood, white blood cells account for less than 1 per cent, so the number available in a small quantity of blood was tiny.

That wasn't the only weakness of the technology. While an electrophoretic image for a traditional blood grouping sample could be produced overnight, an image for an early DNA fingerprint could take weeks or even months to produce. I'd often have police officers calling me for an update on their cases and I'd have to explain that, although the photographic film had been positioned over the sample, removing it too early meant there was a good chance the image might not yet be visible.

Especially in cases where there was only just enough body fluid to perform the test once, this represented a huge risk, as there would be no way to repeat the procedure. On this basis, the prevailing attitude at the laboratory was that the longer the plates could be left, the better the chance of a readable profile being obtained. I'd carefully explain this to the officers concerned and invariably they agreed to wait as long as they possibly could.

From that time, every case we received with limited material became the subject of a dilemma. Should we perform a DNA test in the hope that we got a more specific result, or use the material for blood grouping instead, where the chances of getting a quick result were far higher?

It didn't always work out for the best. There would be times when cases would be thrown out or charges dropped because of the time taken to produce a profile. Frustratingly, the police would not always tell the laboratory when this happened, so I'd sometimes find myself writing up reports and submitting results for cases that had already been heard in court.

Although DNA tests carried out on properly preserved samples taken from suspects were almost always successful, the percentage of successful profiles gained from samples recovered from crime scenes was still low.

In theory, each individual has a unique multi-locus DNA 'fingerprint', but the multiple bands they produce are difficult to read and interpret. Slight variations in the way the electrophoresis gel had run or the samples had been lined up during preparation could make it appear that the resultant bands didn't match when they did, or vice versa. In addition, the results could be ambiguous if fragments were close together, as a particularly dark band on the X-ray film could be caused by several bands being superimposed over one another.

In practical forensic interpretation terms, the ideal was that the samples from the suspect and the crime scene would be processed next to one another so that it would be easier to identify corresponding bands in the two samples. It wasn't always that easy, though, and I'd sometimes find myself using a plastic ruler – one with plenty of flexibility in it – to demonstrate that two bands did in fact line up, even if they didn't appear to at first. However, in terms of the risk of cross-contamination, processing suspect and crime scene samples next to each other was the worst possible scenario.

Despite this, a flurry of court cases was still making its way through the system, and DNA seemed to be fully accepted by both judges and juries on both sides of the Atlantic as little short of infallible.

DNA profiling became part of the armoury of the US judicial system soon after its debut in the UK. Cellmark, the company Alec Jeffreys had helped to set up, opened up a branch in Maryland in 1987, hoping to be the first to offer its services to prosecutors and defenders. It was, however, beaten to the punch by the New York-based Lifecodes Corporation, a paternity-testing company which had launched in 1982 and subsequently moved into the DNA field. Crucially, the company had developed its own version of the Cellmark test – one not covered by the worldwide patent.

Both firms made use of Restriction Fraction Length Polymorphism analysis (the process which looks for repeating stutters in DNA fragments), but while Cellmark used the multi-locus analysis to examine many parts of the DNA sequences in one test, Lifecodes used a single-locus analysis which looks at only one site at a time.

Rather than using a collection of radioactive probes to bind to VNTRs at multiple locations on the DNA, the single-locus probe (SLP) binds to a single location. The process is identical to that of multi-locus probing other than that, once developed, the X-ray film produces just one or two bands. SLP results are easier to read and also have the advantage of using up less forensic material. Mixed body-fluid stains are also less problematic to interpret with this technique. The DNA profiling results can be converted into a series of numbers which are searchable on a database. An SLP database was later constructed in the UK to contain undetected crime stains and profiles of individuals in cases which involved the presentation of DNA evidence – an early forerunner of the current UK National DNA Database.

However, while MLP is a highly discriminating technique and as a result is seen as 'the equivalent of a fingerprint' in terms of uniqueness, an individual SLP result may apply to a far larger proportion of the population. To counter this shortcoming, several SLP analyses are performed sequentially with different single-locus probes until the certainty of a match to a particular individual rises above 99 per cent. The SLP method builds up the 'bar code' in a number of stages and is more suitable for crime stains in which the DNA is often damaged or degraded. Alec Jeffreys quickly acknowledged the advantages of the SLP technique and set about developing one for Cellmark.

Lifecodes performed the tests in the first case in the United States in which a criminal was identified by DNA, that of a rapist named Tommie Lee Andrews, who went on trial in Florida in November 1987. A scientist from Lifecodes testified that semen taken from the victim matched Andrews' DNA, and that his profile would be found in only one in 10 billion individuals. The jury returned a guilty verdict and Andrews was subsequently sentenced to 22 years in prison.

This case was heavily reported in the press, which created a media buzz that was highly favourable to this new technology which was able to identify criminals with 100 per cent accuracy. Other cases quickly followed, with similarly effective results. Judges and juries were clearly impressed.

In New York, when the jury in the case of a murderer and rapist found the defendant guilty largely on the basis of the DNA evidence, Judge Joseph Harris declared the technique to be 'the single greatest advance in the search for truth since the advent of cross-examination'. Even leading defence attorneys were forced to admit they were almost powerless to act in the face of DNA evidence. 'In rape cases, when the semen has been matched with the defendant's and the chance that it came from another person is 33 billion to one, you don't need a jury,' said one.

However, while the MLP and SLP approaches were valid and reliable from a forensic perspective, the techniques had significant limitations. A comparatively large volume of long lengths of intact and high-quality DNA was required for RFLP analysis. These could be broken down by degradation into shorter fragments which could not be analysed by this method. The techniques were laborious and time-consuming and radioactive materials were required for detection. Significant differences in band sizing were regularly observed during collaborative exercises carried out by DNA-testing laboratories throughout Europe. Although measures were taken to standardize as many variables as possible, including buffer system, gel running time and temperature, and DNA visualization method, there was still an uncertainty in the uniformity of interpretation between laboratories.

MLP, in particular, had additional drawbacks for forensic use: a large amount of biological material was required to produce a result; difficulty was encountered in analysing stains from mixed body fluids such as semen and blood together or semen from several individuals in a multiple rape; and the results were unsuitable for recording on a database.

Cellmark and Lifecodes may have made their moves early, but they missed out on the next big development of DNA, which came about with the case of Gary Dotson. By the spring of 1988 Dotson had spent more than a decade in prison, having been convicted in the late 1970s of raping a 16-year-old cook at a fast-food restaurant in a Chicago suburb. The victim said she had been attacked by three men, one of whom had tried to write on her stomach with a broken bottle.

Blood grouping showed Dotson, just 22 at the time of the incident, to be a one-in-ten match to semen found on the woman's underwear. Despite many inconsistencies in the evidence – among other things, the victim stated the rapist was clean-shaven, but Dotson had a full beard and moustache when he was arrested just a few days later; and she claimed to have scratched the chest of her attacker, but Dotson had no injuries – he was found guilty and sentenced to 25 to 50 years in prison.

In early 1985 the victim, Cathleen Crowell, confessed to having fabricated the rape allegation that had sent an innocent man to prison. She said she had invented the story because she feared that her boyfriend at the time had made her pregnant. She thought she needed a cover story if that turned out to be the case as her foster parents had threatened to throw her out if she started having sex. She had inflicted the cuts on her stomach herself and had torn her clothing to fortify the false claim.

Dotson was released on bail, pending a hearing which was due to take place a week later, but the judge dismissed new evidence that sought to discredit the original forensic evidence at the trial and sent Dotson back to prison, calling Crowell's recantation 'less than credible'. It didn't help that blood grouping had failed to link Crowell's then boyfriend to the semen stains left on the underwear she had been wearing during the 'rape'.

Dotson's attorneys immediately petitioned the then governor of Illinois for clemency. After a three-day hearing, the governor denied clemency but commuted Dotson's sentence to time served, allowing him to leave prison the same day. It was a move that led to much confusion. If Dotson was guilty, why was he being released? If he was innocent, why was his conviction being allowed to stand?

Two months later, Dotson was involved in a domestic violence incident and returned to prison. Though the charges were dropped, he had breached his parole, which meant he had to finish serving the remaining 16 years of his rape sentence.

It was around this time that Thomas Breen, Dotson's lead attorney, came across an article in *Newsweek* magazine entitled 'Leaving Holmes in the Dust'. The article reported the discovery of a technique capable of linking criminal suspects to crimes through DNA and described the technique as 'the molecular equivalent of dusting for fingerprints'.

The article explained how the technique had been used to resolve a handful of paternity and immigration cases in Britain and, more dramatically, to link Colin Pitchfork to a double murder. Although Breen learned upon inquiry that DNA had never been used to exonerate anyone already convicted in a criminal case, its potential for doing so seemed obvious.

Alec Jeffreys was contacted and agreed to do the testing

through Cellmark, but the technique he was using at the time was one of those that required a large amount of long lengths of intact and high-quality DNA in order to work. Samples that were too old or degraded – as was the case with the decade-old underwear that formed the key evidence in the Dotson case – simply failed to produce any result.

Although Lifecodes had developed single-locus probing, it still relied on RFLP analysis, so if Jeffreys had been unable to get a result, it was highly unlikely that they would. Dotson's attorneys were just about to give up, when a new player entered the field.

California-based Forensic Science Associates had been developing a new way of identifying individuals from their DNA using a technique first patented by the Cetus Corporation in the mid-1980s known as polymerase chain reaction (PCR). Kary Mullis at Cetus had discovered a way to use PCR to amplify one particular part of the DNA sequence, making it possible to obtain results from samples even when they were old and had degraded.

PCR mimics the natural process by which DNA replicates itself. The DNA sample is heated until the two strands of the double helix separate. When the sample is cooled, each half-strand acts as a template, attracting a mirror image of itself from additional DNA 'bases' placed into the mixture and thus creating an exact copy of each strand. It is this perfect copying that lends itself to use in forensic science.

By using an enzyme called DNA polymerase, obtained from bacteria living in hot springs and geysers, the sample could be put through multiple cycles of heating and cooling to produce millions of copies of the original DNA fragment in a matter of hours.

The area of DNA copied in the system developed by Forensic Science Associates was known as HLA-DQ alpha and occurred

Polymerase Chain Reaction

Genomic DNA

① Heat makes DNA unzip (95°C)

② Primers (~55°C) stick to target gene

Polmerase enzyme adds nucleotides (72°C)

③ Two copies of gene form one strand of DNA

Repeat steps 1-3, 30-40 times

Millions of copies of target gene!!

The PCR process: The DNA is first heated, forcing it to separate into two strands. As it cools, DNA fragments from the polymerase form bonds – G to C and T to A – thus creating two exact copies of the original strand. The mixture is then reheated to separate and duplicate these new strands. The process is repeated until millions of copies of the original have been created.

in one of 21 possible variations. Although this was much less discriminating than the results produced by other forms of DNA profiling, it still meant that the probability of a random match between a suspect and a semen sample ranged from one in seven for the commonest gene type (allele) to one in 100,000 for the least common.

In the Dotson case, the tests revealed that the spermatozoa on the victim's undergarments could not have come from Dotson but could have come from the victim's boyfriend. Dotson's conviction was overturned – the first case in which someone convicted of a crime was exonerated through the use of DNA.

Within the space of a few short months, the introduction

of single-locus probing and the use of PCR meant that DNA techniques could be used more effectively. Not only were the results of tests simplified to make comparisons easier, but the technique could be performed with far less material. While most DNA tests worked best with fresh samples, PCR had the added benefit of producing results from stains that were as much as 15 years old.

Although analysis of small spots of blood remained problematic, previously insignificant quantities of semen proved to be a rich source of genetic material, as each individual sperm head was essentially a packet of DNA, providing an excellent basis for profiling. For this reason, DNA would first rise to prominence within the criminal justice system due to its role in cases of sexual assault. However, the very first case the MPFSL solved as a result of DNA analysis involved a very different bodily secretion.

In December 1988 a 22-year-old photographer named Lorraine Benson was murdered in London. Her body, naked except for a jumper and a pair of socks, was found hidden in brambles beside a footpath in Raynes Park. The rest of her clothing, along with her handbag, was found near by, along with a man's handkerchief.

An autopsy found that Benson had been strangled and that the killer had used a length of rope found near the body, which had left a patterned line around her neck. She had also been punched on the nose and chin and beaten around the head, and there were bite marks on her left hand and arm.

No semen was found on the vaginal and anal swabs taken from the body and none was found on her clothing. Tiny traces of saliva were discovered on swabs taken from her breasts and from a bite mark on her arm but when attempts were made to get results using the ABO system, they were unsuccessful.

At the time, there had been another series of rapes and attempted rapes all within three miles of the area where Benson's body had been discovered, and the lab's DNA unit was working flat out, screening large numbers of suspects to see if a match to the semen samples obtained in the rapes could be found. Following the capture of Colin Pitchfork in the case of the Narborough murders, such mass screenings were becoming increasingly common as a way to eliminate vast numbers of suspects, though the pressure this placed on the facilities within the laboratory were enormous.

Police did not believe that Benson had been murdered by the man responsible for the series of rapes but asked the laboratory to see if there was anything they could do by way of obtaining a DNA profile from the materials that had been recovered from the scene.

The handkerchief had been found a little way from Benson's body and could have been unconnected but was tested just in case. With crime scenes, it's always better to be safe than sorry and to collect and analyse everything, rather than jump to conclusions about whether items belong to the victim or their killer or not.

The handkerchief had a few bloodstains on it, some of which had been mixed with saliva, but the fabric also had spots of a crusty yellow material which was quickly identified as nasal mucus. Mucus within the nasal passages is normally clear, but if the body is fighting an infection, such as a cold, a large number of white blood cells are secreted within it in an attempt to kill off the invading virus. This made the mucus a strong candidate for DNA profiling.

In February 1989, a few weeks after the murder, a 19-year-old known sexual offender named John Dunne was arrested on an unrelated matter. He was questioned intensively but denied knowing anything about Lorraine Benson's death. He offered no objections when asked for specimens of intimate bodily fluids, and samples of his nasal mucus and blood were despatched to the laboratory, together with his dental impressions.

The results were unequivocal. The mucus on the handker-chief indicated an occurrence of one in 1,497,000, which meant that only forty people in the white population of the UK could show such a profile; of that 40, half would be women. In other words, the evidence was that, out of all the white men in Britain, only 20 could have produced that mucus. And included Dunne.

In addition, odontology tests showed that Dunne's teeth were a perfect match for the bite mark on Lorraine Benson's arm. Confronted with the evidence, Dunne admitted to the murder and was sentenced at the Old Bailey to life imprison-ment.

Once it became clear that DNA was going to be a key tool in the future of forensic science, it became hugely important that all reporting officers who would be presenting evidence at court had a strong enough understanding of the process to be able to explain it to a jury.

Many of those working in the general search department were graduates in applied biology, while some of those in the biochemistry department had more specialized degrees. DNA itself fell within the field of molecular genetics and was therefore a subject that had been dealt with in varying depth in different biology disciplines. We had all learned about Watson and Crick and the basis of molecular genetics during the course of our university studies, but our knowledge didn't always go a great deal further. If we were going to be able to do our jobs effectively, we needed to acquire a vast amount of new knowledge.

To this end, the laboratory arranged for us to attend a series of evening classes at King's College. The *quid pro quo* for the college was that King's had recently set up an MSc in forensic science and wanted staff from the laboratory to deliver a few lectures on specialist topics and the practicalities of the work. It was a win–win situation all round and I lectured on the King's MSc course for several years thereafter.

After months of theory, the course culminated in a two-week residential course at Hatfield College which covered the practical side of DNA profiling. Hatfield was chosen because, at the time, it had one of the most advanced practical DNA laboratories in the country.

None of the training cases we worked on were live – of course – but having such hands-on experience meant that all the reporting officers would be able to speak with authority about every aspect of the DNA process in the future, which was just as well. DNA quickly changed everything about the way forensic science was viewed within the criminal justice system. With DNA analysis, the police were suddenly able to eliminate suspects rapidly, focusing far more of their time and attention on those who could not be eliminated.

DNA also changed the way that defendants in rape cases behaved. Two common defences used to be to say that the rape simply had not taken place or that some other person had been responsible. With DNA providing near 100 per cent certainty of the source of semen found after intercourse, defendants were left only with the defence of consent.

DNA also upped the ante so far as evidence was concerned. In addition to fibres evidence, Everald Irons had been convicted on the basis of a traditional semen grouping result that would be found among just one in 300 of the Afro-Caribbean population. Although that theoretically meant there would be at least 1,500 other Londoners who would produce a similar result, this was considered to be a good result at that time and one the court was more than happy to accept.

Once DNA came in, the chances of any two people sharing a profile increased first to the hundreds of thousands – as in the Lorraine Benson case – then to the millions and, finally, the billions. Suddenly, the courts – or, to be more specific, juries – were no longer impressed with numbers like one in 300. Unless you had a string of zeros at the end of your result, your evidence was far more likely to be disputed or even ignored.

But the numbers have only ever been part of the story. A match probability of one in 50 of the population means that a result could apply to hundreds of thousands of others, but it fails to take account of other factors, such as proximity to the location of the crime, age of the offender (some of those potential matches would be under 18 or over 65), and so on.

No sooner had DNA been accepted by the courts than the first of many challenges against it began to emerge. The key issue was that, because DNA was so new, the idea that no two people in the world could share the same profile was the result of a statistical assumption, not one that could be demonstrated empirically.

In the case of blood grouping using the ABO and other systems, there were extensive tables showing the proportion of the population with a particular blood group or factor. These tables were based on extensive testing that had been going on for decades. Many thousands of tests had been carried out and, as a result, the data was deemed to be fully reliable.

When DNA first appeared on the scene, only a few hundred analyses had ever been carried out – indeed, the first statistical extrapolations were said to be based on '44 people living in the Leicestershire area'. Although the numbers were growing rapidly, it would be years before the figures were high enough to bury any lingering doubts.

Despite this, police had begun to charge burglars purely on

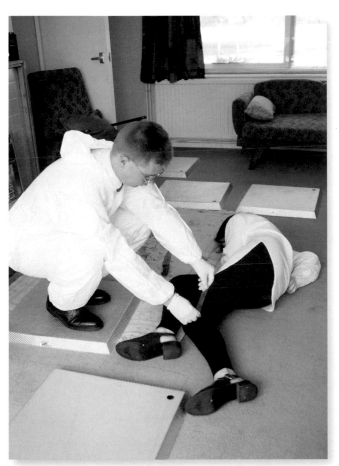

ABOVE: A Kent Constabulary crime scene van (*c*. 1991). Standard equipment included a scene tent, generator and arc lights, sieves and digging equipment for fires and exhumations.

LEFT: A SOCO instructor uses a student to demonstrate the correct method of collecting extraneous material from a body. Note the use of treadplates to protect the crime scene.

My original sketches of the Islington beheading murder scene featured in chapter one. Drawings allow the distribution of key blood patterns to be seen more clearly than in a photograph.

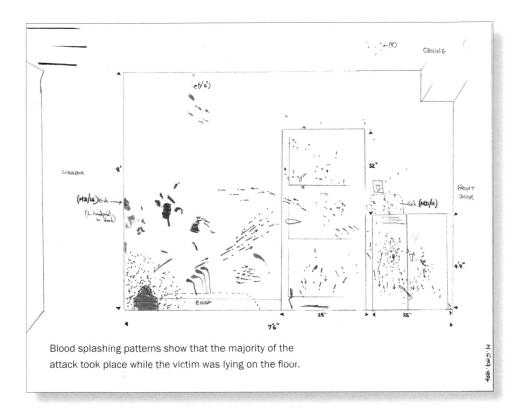

Blood splashing patterns show that the majority of the
attack took place while the victim was lying on the floor.

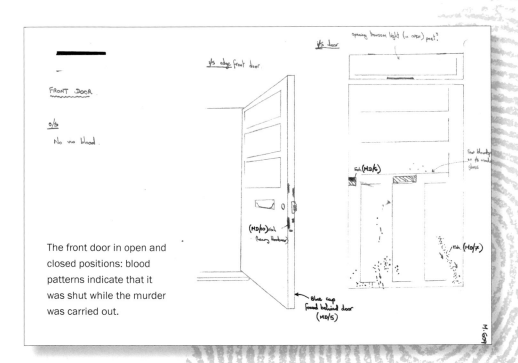

The front door in open and
closed positions: blood
patterns indicate that it
was shut while the murder
was carried out.

Polaroid photographs taken by me of the same murder scene depict the body lying in the hallway. Drag marks in the blood (**below**) indicate that the body was moved to let the assailant escape through the front door.

ABOVE: The scene viewed from inside the flat looking out towards the front door.

RIGHT: Blood splashes originating from the victim's head.

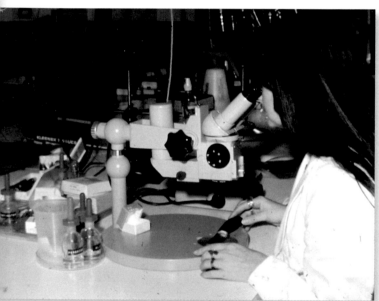

ABOVE: Performing a Kastle-Meyer test to identify whether blood is present in a stain. A red colour indicates presence of blood.

LEFT: The blade of a knife is examined under a low-power microscope.

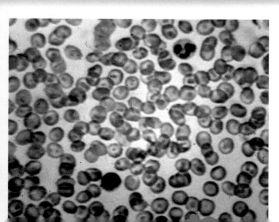

LEFT: Human blood seen under high magnification. The white blood cells have been stained blue to make them easier to see.

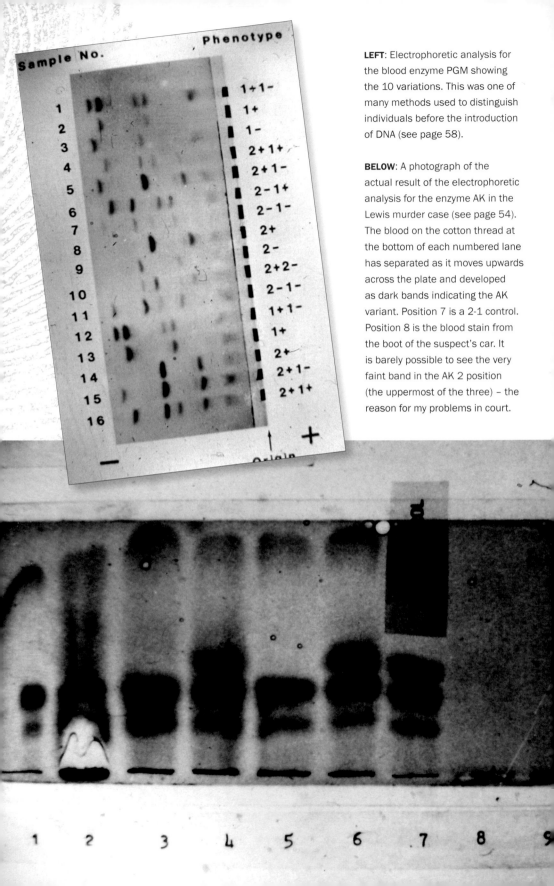

Sample No.   **Phenotype**

1   1+1−
2   1+
3   1−
4   2+1+
5   2+1−
6   2−1+
7   2−1−
8   2+
9   2−
10   2+2−
11   2−1−
12   1+1−
13   1+
14   2+
15   2+1−
16   2+1+

−   ↑   +
Origin

LEFT: Electrophoretic analysis for the blood enzyme PGM showing the 10 variations. This was one of many methods used to distinguish individuals before the introduction of DNA (see page 58).

BELOW: A photograph of the actual result of the electrophoretic analysis for the enzyme AK in the Lewis murder case (see page 54). The blood on the cotton thread at the bottom of each numbered lane has separated as it moves upwards across the plate and developed as dark bands indicating the AK variant. Position 7 is a 2-1 control. Position 8 is the blood stain from the boot of the suspect's car. It is barely possible to see the very faint band in the AK 2 position (the uppermost of the three) – the reason for my problems in court.

1   2   3   4   5   6   7   8   9

Adding a sample by hand into an early DNA sequencer. The separation of the DNA takes place vertically downwards with the detection occurring at the bottom of the sequencer.

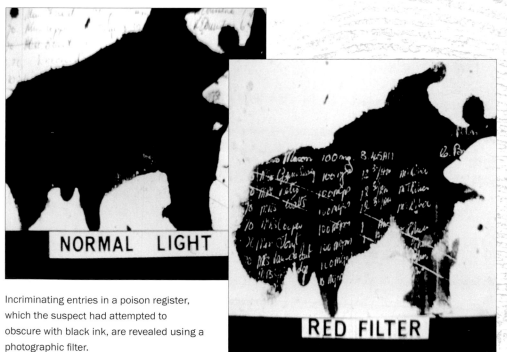

NORMAL LIGHT

RED FILTER

Incriminating entries in a poison register, which the suspect had attempted to obscure with black ink, are revealed using a photographic filter.

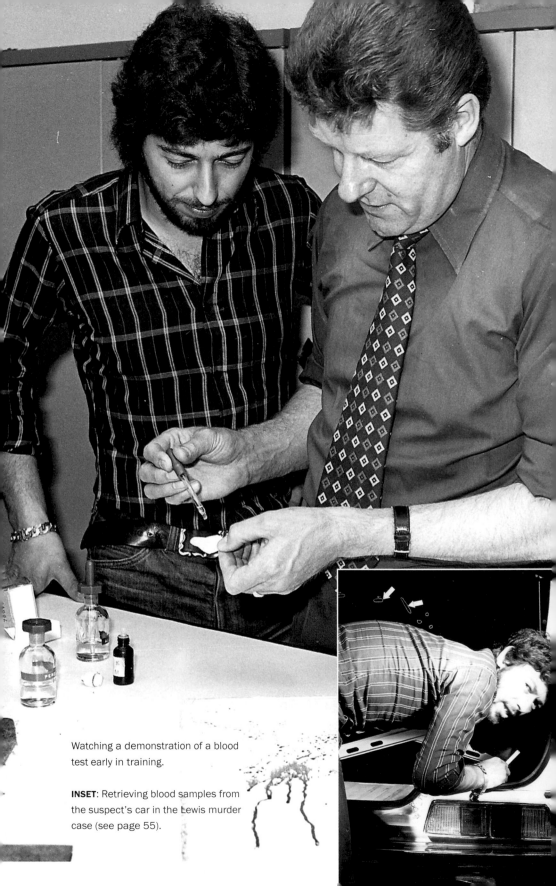

Watching a demonstration of a blood test early in training.

INSET: Retrieving blood samples from the suspect's car in the Lewis murder case (see page 55).

the basis of DNA evidence found at crime scenes, rather than carrying out any further investigation.

The Crown Prosecution Service (CPS), however, was unhappy with this and insisted that there had to be additional evidence, otherwise it would not proceed with the case.

9

THE FORENSIC SCIENCE LABORATORY WAS OWNED BY THE Metropolitan Police, but they were not our only clients. We also carried out forensic work for police forces in the home counties, the Ministry of Defence, Intelligence Services and numerous British protectorates on behalf of the Foreign and Commonwealth Office.

There was a fairly steady stream of overseas cases. Some reporting officers had families or, for other reasons, preferred not to go abroad, while others, such as myself, were happy to have the opportunity to travel. The overseas cases were handed out on a simple rotation basis so that everyone who was willing and able to do them would have an equal chance.

However, just because you were assigned an overseas case didn't mean you would be given the chance to visit the country in which it had occurred. Not all cases went to court and it was perfectly possible to spend time doing analysis or writing reports but never be called to make a personal appearance.

One of my very first overseas cases involved a sexual assault in Diego Garcia, a coral atoll just south of the equator which is part of the British Indian Ocean Territory. Depopulated in the 1960s, the foot-shaped island now serves as a joint military facility for the United States and Great Britain.

With no tourists allowed, no commercial flights and permits granted only to yachts requiring safe passage, a visit to Diego Garcia would have provided a once-in-a-lifetime chance to visit a country that only a handful of people will ever see. I kept my fingers crossed but, sadly for me, no trip overseas was required and my name moved to the back of the queue to await the next assignment abroad.

A few months later, I worked on a case involving an army chef who had been involved in the stabbing of another soldier, and I ended up travelling to Osnabrück in Germany to give evidence at the court martial.

It was while I was there that I was taken out by a group from the sergeants' mess. We visited a bar and, having managed to pass some drinking initiation test, I was invited to join the prestigious 'Jägermeister Club'. My name was engraved on a mayoral chain worn by the club president – I still get invitations to the annual formal dinner of asparagus and Jägermeister. Dropping me back afterwards, as they weren't allowed in, the sergeants kicked me off the tailgate of their military Land Rover into the gravel outside the officers' mess where I was staying. My appearance at the court martial the next day was fortunately postponed to the afternoon.

The destination may not have been as glamorous as the Indian Ocean, but the court martial itself was every bit as elaborate, with all those concerned in full dress uniform. When asked to inspect one of the exhibits to confirm it was the same one that I had performed my tests on, I had trouble removing it from the sealed plastic bag it was contained in. 'Perhaps this will help,' said the Judge Advocate, removing his three-foot-long ceremonial sword from its scabbard and passing it to me across the room.

Then, just before Christmas 1988, evidence from another overseas case arrived at the laboratory and, as I had once again reached the position where I was next in line, I was called in by my boss and offered it.

The case was that of two highly regarded American psycho-therapists, Susan Galvin and Martha Marie Alsup, who had been raped and murdered on the Caribbean island of Anguilla while on holiday.

Made into a British colony in 1650, Anguilla had been incorporated into a single colony along with the islands of St Christopher (commonly known as St Kitts) and Nevis in 1825. This federation collapsed in 1962. The creation of the Associated State of St Kitts-Nevis-Anguilla by the British government in February 1967 – without the wish of the people – sparked the Anguilla Revolution.

On 30 May 1967 Anguillans marched on their police headquarters in The Valley (the capital) and expelled the 13-man contingent of St Kitts policemen. This 'act of revolution' was dealt with in typically heavy-handed fashion by the British govern-ment sending an invasion force of heavily armed paratroopers to the island. They were greeted by union-flag-waving welcoming parties of delighted Anguillans. The revolution was historically short-lived and bloodshed-free. However, as a result, the island was henceforth policed by a contingent from London's finest: the Metropolitan Police – not a hardship post!

The Royal Anguillan Police Service was founded in 1972 and, today, around 40 officers are responsible for the day-to-day policing of the 13,000 residents of the island, but when-ever specialist help is needed, detectives from Scotland Yard are brought in.

In the immediate aftermath of the murders of the two women in 1988, a local youth, Andy Otto, was arrested in connection with the deaths. The bodies of the women had been found on an isolated beach known as Windward Point, which Otto, who lived near by, considered to be his personal territory.

Serious and violent crime was rare on the island – one sexual assault every five years, one murder every decade. Otto had, however, been responsible for a one-man crime spree. He had been accused of raping a tourist in September 1988, but no charges were brought because the victim said she would not return to Anguilla to testify at a trial. The following month, he was held in jail for a short time, having been charged with raping a woman tourist at knife-point in October, but was then released on bail.

In November, just two weeks before the murders of Alsup and Galvin, Otto had been caught allegedly attacking another female tourist on a beach. Naked, he was chased by a group of fishermen, but he managed to slip away.

Galvin and Alsup, who lived as a couple in Watertown, Massachusetts, were both raped, beaten with rocks, stabbed and left to bleed to death. Arrested within a few hours of the killings, Otto initially confessed to robbing the women and told police where he had stashed their money and other personal belongings, but he denied attacking them.

From my point of view, the case was just about as straightforward as it was possible for any case to be. The investigating team collected a number of swabs from the victims' bodies, along with blood samples from the suspect, and sent them in for comparison.

As this was a rape case, I would be focusing my attention on looking for semen. I carried out presumptive tests on the swabs, which indicated that it was present, and then cut a small piece out of each swab and examined each one under the microscope. This also confirmed that sperm were present. After that, all I had to do was package up the samples, put them in the freezer and submit a request for DNA comparisons to be run as quickly as possible.

Although it was urgent, the DNA technique available at the time meant that it would take weeks before any results were available for viewing. The actual tests were carried out by my colleagues in the DNA unit rather than by me, and my only stipulation was to make sure that both the blood and the semen samples were run on the same day.

Tiny variations in the electrical conditions and strength of the buffer solutions used could mean that, if you ran the tests on separate days, the bands might still match, but perhaps not with the same level of precision. Although this would not cause any doubt about the veracity of the findings from a scientific point of view, it would be all too easy for a defence barrister to question the results.

Several weeks went by before my DNA colleagues and I had a conversation and agreed that the plates would be ready for inspection. We knew that the longer we left it, the clearer the results were likely to be and that, if we viewed the samples too early, we'd be forced to start all over again. As it was, the results were absolutely conclusive. Semen had been found inside both women and DNA analysis showed it was a match to Andy Otto. There was no question of contamination, no question of the fluids having found their way inside the women in any other way. It was a straightforward case, from a scientific point of view at least.

My statement was only a couple of pages, most of which was taken up with providing information about the chain of custody of the samples and the continuity of my analyses, to prove that the material tested definitely came from the women and had been compared with samples from Otto himself.

Normally, the plates would be photographed with a simple Polaroid camera, but because this case was going abroad I decided to get someone from the photographic department to take higher-quality images.

A statement containing the results of the analyses was sent off to the officer in charge of the case and, a few weeks later, I received a request to travel to Anguilla in order to testify at the trial. It would be the first time that DNA evidence in a UK police investigation had ever been presented to a court outside the UK and, thanks to a simple twist of fate, I was going to be the person to present it. Although the case and the evidence were straightforward enough, this was going to be a high-pressure assignment for all sorts of reasons.

By then, the decision had been made that, despite his youth, Andy Otto should be tried as an adult. Feelings were running high. The island was heavily dependent on tourism for income, and the murder of two American tourists, plus the other attacks Otto was suspected of, had the potential to make a huge dent in the local economy.

There were other considerations too. Although Anguilla was a British dependency, the island had its own laws and punishments on the statute books, including the death penalty for murder. In reality, the penalty was usually commuted to life, but there had been cases of Anguillans being transported to nearby St Kitts and hanged there. Also, with so much concern about the effect Otto's actions would have on the economy, the local papers were full of calls for him to face the full penalty of the law.

With Otto denying rape and murder, the key evidence against him was the DNA. Another strong piece of evidence had been lost a few months earlier. Either before or after the rape, Otto had tried to break into the car the women had been using to get around the island, forcing open the door by inserting the tip of his knife into the lock.

The knife had broken, leaving the tip inside. As the rest of the knife itself had been found in Otto's possession, it would have been easy to match the tip to the rest of the blade to produce a 'mechanical fit' to show beyond any doubt that Otto had been the one attempting to steal the car. (See page 5 of the second picture section.)

Unfortunately, although a sharp-eyed police officer had spotted the tip in the lock, he had neglected to place a piece of tape over it to protect it while the vehicle was hauled on to a low-loader to be transported to the police station for further examination. By the time the car had arrived, the tip had fallen out and was nowhere to be found.

Without the DNA, the case against Otto was weak. In many ways, the entire case was going to rest on my testimony.

Usually, forensic scientists are called in towards the end of a court case but, for Otto, the Attorney General wanted me there at the beginning of the trial so that I would be able to brief the prosecution counsel on the DNA evidence I'd be presenting, as it was all so new. This meant that I would be in Anguilla for at least 10 days. That was exciting enough, but this particular assignment came with an additional perk.

The Metropolitan Police Laboratory worked according to civil service rules and regulations and, as the ambient temperature in Anguilla was going to be over a certain level during my time there, I would be entitled to a tropical suit, which the laboratory would purchase on my behalf. I immediately had visions of some multi-pocketed khaki outfit, complete with pith helmet, of the sort that David Attenborough would wear while out on safari. I couldn't have been more wrong.

The suit would be a standard smart grey pinstripe made of two layers of a special ultra-light weave devised by Piccadilly tailors Airey and Wheeler, who supplied many such outfits to the civil service, the Foreign Office and the military. You couldn't tell it apart from a regular suit in most circumstances but if you held it up to the light you could see right through it. The lightness also meant that the wind blew right through the material, helping to keep you cool. I was given the option of short or long trousers.

As I knew I'd have to spend a lot of time explaining DNA analysis to the jury, I packed one suitcase with all the visual aids and explanatory material I thought I'd need. It turned out to be bigger than the one I had for my clothes. My reasoning was simple: I didn't know exactly what they were going to ask or how detailed an explanation they would require. I wasn't going to be in a position to send someone back to the laboratory to grab something I'd forgotten, so it was best to have absolutely everything with me.

I flew out via New York, put on my tropical suit and went to see the Attorney General. Although there was lots of tourism on the island, most visitors stayed in one of a small number of resorts, so hire cars were few and far between. The Scotland Yard detectives had taken the only vehicle the police could spare, so at first I was stuck, but then help came in the form of the local undertaker. He informed me that he had a spare car I'd be able to use, but that if anyone died, he'd need to borrow it back. Thankfully, it wasn't a hearse but the car used by relatives when following the funeral procession.

After explaining the significance of the evidence to the Attorney General, I was free to wander about, as I wouldn't be required to testify until towards the end of the prosecution evidence, which was going to be at least a week away.

Driving around the island, it soon became clear that the trial was huge news and that everyone was aware of exactly why I and the rest of the police team were there. 'You are going to hang him, aren't you?' was a common question.

It may be the capital of Anguilla, but The Valley, with a population of just over a thousand, was barely a town – more a large village. At the centre was a compound surrounded by a 10-foot-high chain fence which housed the court-house, the environmental health office and a handful of other government buildings.

With so little crime on the island, the court-house had many other uses. During the night it was used to house chickens, so, each morning, before the judge and jury were able to enter, the chickens had to be evicted and any mess they had left behind swept up. During the lunchtime recess, the court-house would become the registry building, leading to the bizarre situation of having couples taking their vows in virtually the same spot where an accused murderer might be listening to the case against him.

A large, open-plan building, there was nowhere inside for witnesses to wait until they were called – they can't wait inside the court, as they aren't allowed to hear evidence from others until they have been dismissed. The compromise was to install a seat under a tree in the court-house grounds, which had become known as the 'witness tree'. Although not ideal – even for those of us blessed with tropical suits – it at least afforded some protection from the blazing sun while you were waiting to be called.

The main road passed through the middle of the compound but whenever the court was in session the gates would be closed and drivers would have to take the long way round. Few bothered. The fence adjacent to the gates had been trampled down, so it was possible to turn a few feet to the right and carry on through the compound regardless. Even though every single driver took advantage of this short cut, the staff at the court-house still made a big show of coming out and closing – or opening – the gates in accordance with court times.

During the first week, the whole team had little to do apart from sort through the many photographs and exhibits that would be needed once the trial proper started. The best place to do this, we decided, was while sitting in the giant jacuzzi on the deck at the rear of our hotel. What the tourists thought of this group of Englishmen passing each other copies of grisly autopsy photographs, statements and other items, we'll never know.

At the end of that week the police team was required in court, so I was left to my own devices. By then my girlfriend had taken some time off and flown out to join me so we could enjoy a break together while I waited to appear.

Each day Detective Superintendent Roy Ramm, the officer in charge of the case, would let me know how things were going, but it was clear that I wasn't going to be called at all during the whole of the second week. By the third week, a similar picture was emerging. On the Wednesday evening, Roy informed me that I definitely would not be required the following day but would be the day after that, so it was my last bit of freedom.

For a change, my girlfriend and I decided to take a boat to a nearby island, which had a basic bar and nothing else but was said to be a great place for a day trip. We found a local with a boat who was willing to take us there for a modest fee and spent the afternoon slowly getting sloshed on high-proof rum before making our way back.

When I returned to my hotel room, I saw a letter on my

bed from Roy. It turned out that the court had finished its other business earlier than expected and was now waiting on me to give evidence. With no mobile phone or any other way to get hold of me, Roy had had no choice but to leave a note. Unfortunately, by then I was in no fit state to do anything, let alone give evidence in a murder trial. The proof of this was that, as I made my way out of the hotel to find Roy, I fell into the pool – fully clothed.

The next day, with a slight hangover, I spent a brief spell under the witness tree before making my way into the court-room. The judge was a short woman from Barbados with an enormous blue-rinse Afro who had been flown in especially for the trial. Both the prosecution and the defence barrister were from England.

We began with my evidence in chief, going into some detail about how DNA is extracted from samples and giving every detail of the process employed to produce the bands on the photographic plates that were used to determine whether a suspect is a match. Once that was complete, I braced myself for the start of the cross-examination. Knowing how much was resting on the evidence I was presenting, I expected the barrister to begin a long, highly detailed probe into the validity of my results.

'Mr Silverman, did you carry out these tests yourself?'

I didn't even have a chance to answer. The judge smashed her gavel on the bench, stood up and addressed the defence barrister angrily. 'This witness has already explained that these tests take weeks to carry out. Do you think he has nothing better to do than sit around and wait for the results of these tests? Of course he didn't do them himself. Now, if you don't have any sensible questions to ask, I suggest you sit down and shut up!'

The barrister stood for a moment, then sat down. 'No further questions, Your Honour,' he mumbled.

Unlike the British court system, the jury in Anguilla was allowed to ask questions of their own directly to the witness rather than simply by passing a note to the judge. The foreman asked how much confidence I had in the results and I explained that the chance of two people sharing Otto's DNA profile was hundreds of thousands to one.

Once I'd answered the question to his satisfaction, he informed the judge that there were no further questions and I was allowed to leave. I'd spent several hours in the witness box giving my evidence in chief, but my cross-examination had lasted just a matter of minutes.

I flew back to the UK the very next day and, while I was in the air, the trial concluded. The jury took just two hours to reach their verdict, finding Otto guilty on both counts of murder. He was sentenced to death, but the penalty was immediately commuted to life with no possibility of parole.

It seemed that Otto was destined to die in prison, but his conviction turned out not to be the end of the story. In December 2004 he briefly escaped, along with another convicted killer, making it all the way to the nearby island of St Maarten before being recaptured. In 2011 he successfully appealed his sentence, based on his age at the time of the original trial, and was released. He chose to remain on the island. Less than a year later, he disappeared overboard during a fishing trip. His lifeless body washed up on shore the following day, close to the beach where he had murdered the two women many years before.

10

IT WAS CLEAR THAT DNA WAS FAST BECOMING THE LYNCHPIN of the forensic science contribution to the criminal justice system. Despite this, the Home Office Forensic Science Service, the organization responsible for performing analyses on behalf of every police force in England and Wales, with the sole exception of the Met, was in danger of losing its status as world leader in the field.

The key issue was funding. The rapid growth of DNA testing required new equipment and training, and DNA testing cost significantly more than previous analyses. At the same time, the technique had proved so useful in everything from murder and rape, to burglary, car theft and criminal damage that the number of items being submitted for DNA analysis was growing rapidly.

Prior to 1989, police forces contributed a set proportion of their budgets, based on their police staffing levels, which entirely funded the HOFSS. Funding of the HOFSS therefore didn't depend on use or how many items each police force submitted.

In February 1989, a report by the Home Affairs Select

Committee found that, once the envy of the world, the HOFSS was desperately underfunded and its staff were overworked and suffering from rock-bottom morale.

The report exposed an alarming deterioration in the service, which it blamed on persistent funding shortfalls. Between 1982 and 1988 expenditure within the HOFSS had been increased by just 9 per cent in real terms, in spite of the soaring crime rate and greater demand for its services. The annual budget of the HOFSS represented a mere 0.3 per cent of police expenditure. The result was that, although the quality of work remained high, the HOFSS was having to turn away thousands of less serious cases referred to it by the police, who were becoming increasingly dissatisfied as a result.

The Home Affairs Select Committee recommended that the HOFSS be expanded and ultimately given executive agency status, meaning it would have the ability to directly charge the police for its services. The MPs also called for radical change in the way the HOFSS was funded so that police forces paid for it according to use, on a job-by-job basis.

In a separate and unanimous recommendation within the same report, MPs also called for legislation to permit the 'genetic fingerprinting' of convicted criminals as, at the time, and unlike the taking of fingerprints, DNA samples could be taken only if a suspect consented.

It was clear that this second recommendation would only further increase the demand on the HOFSS and that, whatever happened, the organization would have to change fundamentally in the way it operated.

In August 1989, just a couple of weeks after I'd returned to the UK after testifying in the Otto case, a 30-year-old woman living close to Plumstead Common was viciously raped in her bedroom while her two young children were downstairs.

Wearing a mask and carrying a Stanley knife, the rapist had forced his way in through the rear door – which had been left open to allow the family cat access – dragged the woman upstairs and made the children wait in the living room while he gagged his victim and carried out the assault before making his escape.

The man responsible for the rape was Robert Napper. At the time, the crime was just another entry in the capital's crimes statistics, but in the years that would follow Napper's crimes would be a major contributing factor to the downfall of the HOFSS (by then known as the Forensic Science Service) itself. This in turn would shake the world of forensic science to its core.

Napper was born in February 1966 and brought up in Plumstead, south-east London. During his first 10 years of life, he witnessed much domestic violence, and when his parents divorced all four children were placed in temporary foster care and underwent psychiatric treatment.

When Napper was 12 years old a family friend assaulted him on a camping holiday. The incident marked the start of a descent into violence. Napper became introverted, obsessively tidy and reclusive after the assault. He would emerge from his bedroom only to bully his brothers and spy on his sister when she was dressing. On one occasion, he shot his younger brother in the face with an air pistol following an argument. Napper would remain in treatment at the Maudsley Hospital until he was 16 years old, when he left school, but he remained living at home until the age of 21, working in a series of menial jobs. That same year, he chalked up his first criminal conviction – a conditional discharge for carrying an air gun in a public place. He was ordered to pay costs of £10.

The rape of the woman in the house close to Plumstead

Common was the first sexual assault Napper had ever committed. The usual swabs were taken from the victim and
submitted to the Met lab a few days later. With sperm being
such a rich source of DNA, the chances were high that a profile
would be successfully obtained. But fate intervened.

That same day, two other DNA samples from unconnected
cases somehow became transposed during an early phase of the
analysis. I wasn't involved in any of the procedures myself, and
it wasn't clear how the mix-up had happened or how extensive
the problem was, but it was clear that the evidence continuity
of these samples had been compromised.

As a precaution against the potential for incorrect results, all
DNA samples being processed in the laboratory in that batch
were destroyed and the DNA analyses repeated. There was
no way of knowing exactly what opportunities were lost as a
result. In some cases there was enough DNA available for new
analyses to be carried out; in others, requests could be made for
the police to provide or collect new samples. In the Plumstead
rape case, the remaining material was, at the time, insufficient
for another attempt at obtaining a profile.

Napper, however, had no idea of his lucky escape, and in the
days that followed he seemed to become filled with remorse at
what he had done to his victim. He went missing for several
days and tried, unsuccessfully, to kill himself by taking an
overdose of pills.

When his mother asked why he had tried to take his own
life, he confessed that some men were after him because he had
raped a woman on Plumstead Common. All too aware of her
son's capacity for violence, Pauline Napper immediately called
the police to report what she had been told. She also told his
psychiatrist.

With no address and no date to go on, police failed to make
the link between the mother's report and the August rape –

after all, the information they had received stated that the rape had taken place on the common, not in a house close by – so no further action was taken.

Once again, fate had intervened to keep Napper at liberty. Unable to cope with the monster she feared her son was turning into and unable to convince the police to do anything about it, Pauline cut off all contact with him, burning every photograph of him she owned.

Although they were competing for the same business, Cellmark and Lifecodes often joined forces to promote DNA profiling within the criminal justice system. However, behind the scenes, the two companies were desperately trying to outdo one another.

Eager to keep their proprietary products and processes under wraps, they failed to follow the standard methods for testing and validating new scientific methods – publication and peer review. These would normally be followed by replication in multiple labs and evaluation under exacting conditions. Instead, both labs worked in secret, rushing to get a return on the huge investments both had made. But just as DNA appeared to be shifting ever closer to the centre of the forensic science universe, the first real challenge to the validity of the results it produced was launched on the other side of the Atlantic.

Cellmark and Lifecodes were, in essence, vying for position in court-rooms, well aware that publicity from successful cases would make it easier to license their procedures and sell their proprietary reagents and materials to crime labs across the country and beyond.

As DNA testing spread, more labs joined the market, despite the lack of agreed systems to ensure quality control or the presence of any validation studies. It was only a matter of time before a crack appeared somewhere.

When New York police were called to the Bronx apartment of seven months pregnant Vilma Ponce, they found her dead body in the living room. She had been punctured by more than sixty stab wounds. Her two-year-old daughter, Natasha, lay in the bathroom, murdered in a similarly brutal way. The principal suspect was José Castro, a janitor in a nearby apartment building. Castro denied having anything to do with the murders but, while he was being interviewed, a detective noticed what appeared to be dried blood on Castro's watch and asked if he could take it away to be examined.

The watch, along with blood samples from Castro and the two victims, was sent to the Lifecodes Corporation. The blood on the watch was found to be a match for that of Vilma Ponce and the chance of the sample having come from another member of the Hispanic community was more than one in 189 million.

Castro's lawyer challenged the test results and produced several expert witnesses, who testified that the laboratory had not followed accepted evidence-continuity procedures. Although the DNA evidence was correct, it could not be presented to the jury because the laboratory's procedures for handling such evidence were deemed to be too lax.

The tests were carried out in the early days of DNA and, in the rush to have the case ready to take to court as soon as possible, Lifecodes failed to use accepted scientific techniques in reaching their results. The quality of the data it produced was poor and the company did not even follow its own procedures for interpreting it.

While not questioning the basic science behind DNA, expert witnesses began to cast doubt on the scientific techniques and laboratory procedures used. They challenged the laboratory's practice of declaring matches between DNA samples based only on visual inspection of bar-code patterns rather than by means of a more objective method. They also criticized the company's failure to discount two bands of DNA in the case as non-human contaminants.

Four of the scientists – two for the defence and two for the prosecution – even took the highly unusual step of meeting outside the court and issuing a statement that the data in the case was unreliable, stating, 'the DNA data in this case are not scientifically reliable enough to support the assertion that the (blood) samples . . . do or do not match. If these data were submitted to a peer-reviewed journal in support of a conclusion, they would not be accepted.'

'Although DNA fingerprinting clearly offers tremendous potential as a forensic tool, the rush to court has obscured two critical points,' wrote Eric Lander, a geneticist at the Whitehead Institute for Biomedical Research in Massachusetts in a detailed critique of the Lifecodes laboratory procedures published in *Nature* magazine. 'First, DNA fingerprinting is far more technically demanding than DNA diagnostics [used to determine family relationships or detect inherited diseases]; and second, the scientific community has not yet agreed on standards that ensure the reliability of the evidence.'

Following testimony at a 12-week pre-trial hearing, the judge in the case eventually ruled that, while he would accept the exclusionary evidence that the blood on the watch did not come from Castro, he would not accept the inclusionary tests suggesting that the blood belonged to Ponce.

Up until that point, there had been more than 100 cases in the USA in which DNA evidence had met with little or no resistance, but now, for the very first time, defence lawyers had managed to have it excluded. With no DNA evidence allowed, Castro appeared to be on his way to acquittal – when he suddenly confessed to the crime.

Despite this, the Castro case showed that DNA evidence was

potentially fallible and, as a result, the National Association of Criminal Defense Lawyers set up a DNA task force in the autumn of 1989. The task force initially tried to re-open all the convictions based on evidence tested by Lifecodes. In the end, this never happened, but it marked the start of an unprecedented attack on the validity of DNA evidence on both sides of the Atlantic and led both Cellmark and Lifecodes to introduce new protocols and methodologies to address the issues raised by the case.

The UK government moved quickly to defend DNA fingerprinting in Britain. Viv Emerson, a senior Home Office forensic scientist, said cases in which the technique had been shown to fail reflected the need for uniform laboratory procedures and it was vital that forensic scientists interpreted DNA evidence accurately. Emerson pointed out that, unlike the American cases at the heart of the dispute, such measures were already in place in Britain through standardized procedures, cross-checks and quality controls.

The HOFSS also acted as adviser on forensic science matters to the Home Secretary. The HOFSS had a responsibility for ensuring that all of their scientific techniques were properly validated for use in the criminal justice system in England and Wales. The service had the beginnings of what was to become a complex, if unwieldy, quality-management system, with standard operating procedures for all DNA techniques, validation procedures and interpretation guides, and a comprehensive programme of proficiency testing and inter-laboratory 'blind' and declared trials. In addition, forensic scientists, whether DNA reporting officers or DNA analysts, were trained to an accepted standard and competency. There was also an Inter(national) Laboratory Accreditation Cooperation (ILAC) committee dedicated to DNA standards in forensic science.

Yet the previous year, 26-year-old Bryan Telford from

Swindon had spent seven weeks in prison on a rape charge after voluntarily giving a sample as part of a mass screening. He was released after it was discovered that his sample had incorrectly been identified as a match as the result of a human transcription error.

Also, as the Castro case was taking place, former police officer Brian Kelly was being convicted of rape in a court in Scotland, largely on the basis of DNA evidence he, too, had submitted as part of a screening programme.

Kelly was not identified by the victim of the rape, despite the fact that she knew him well, and an initial analysis carried out by the HOFSS cleared him of involvement in the case. Months later, new analyses carried out by Cellmark indicated that he was the source of semen stains on the victim's dressing gown. He was jailed for six years, but his conviction would later be overturned when it was shown that the techniques used by the lab concerned were faulty. During the DNA analysis, Kelly's reference sample and the crime scene samples had been placed into the gel alongside one another, allowing material from one lane to seep into the other and produce a false positive result. Lab procedures were altered as a result to prevent this from happening in the future.

Human error and poor laboratory procedures were two issues, but there was also the question of the statistical basis used to produce the figures that so impressed juries in such cases. Two prominent American geneticists published an article in *Science*, the leading US research journal, arguing that the DNA analysis that was being used in criminal trials at the time should be inadmissible. They said it was far more likely than previously believed that samples from two people could produce indistinguishable genetic profiles. In one of the cases cited, the chances of error were said to be one in 150 million, whereas the true figure was believed to be closer to just one in 256.

And so the 'Great Debate' began. Population geneticists claimed that only they could correctly navigate the statistical minefield of DNA evidence; statisticians claimed it as their bailiwick; and forensic scientists seized upon it for themselves.

Soon after, Old Bailey judge Mr Justice Alliott called for DNA evidence to be treated with more scepticism until further research had been carried out. Explaining that he had considered DNA profiling to be 'wonder' evidence until he had read material prepared for the defence of an alleged rapist, the judge said that he was astounded by the extremely small samples used as a basis for offering 'very long odds' against defendants.

Alec Jeffreys stepped into the row, admitting that the figures were based on a sample of 1,000 people chosen at random and that this was the basis from which odds of millions to one against an accused being mistakenly identified as the person from whom the blood or semen traces originated. He claimed that any problems stemmed from poor laboratory practices, not from the science itself.

Ever since forensic science was first used as evidence in a court of law, there has been a problem with the presentation of the testimony – not with the science (although that, too, is open to debate in some areas), but with what the results actually mean within the context of a particular set of circumstances: their significance. It is this 'grey area' which allows both prosecution and defence to examine the same evidence or testimony and claim that the exact same results or statements support their opposing views. It is one of the roles of the expert witness to help the court to understand the significance of the scientific evidence – and therein lies the problem.

Take handwriting comparison as an example. A document examiner of many years' experience identifies that the flourishes of the pen strokes on many of the letters of a particular document bear a resemblance to the control handwriting taken from a suspect. In the expert's long experience, such similarities indicate that the suspect is probably responsible for the handwriting in question. *Probably*. Is that the same as 'highly likely'? Or nearer 51 per cent?

Good expert witnesses always ensure that any statement they make in court can be defended during cross-examination. To this end, they avoid absolute words and phrases such as 'definitely' and instead rely on verbal indicators – 'extremely likely' to 'extremely unlikely', for example. So whereas the expert might feel '99.9 per cent sure' when his testimony says 'extremely likely' he cannot give this figure – and the figure may be rather different with another document examiner, and understood to be far from 'definite' by members of the jury or the investigating police officer.

I have been told by some investigating officers that they 'bin' forensic statements on handwriting if the expert has said only that the evidence is that the document is 'highly likely' to have been written by the suspect – something about which the expert may have been 95 per cent sure.

With the advent of DNA, the significance of the scientific evidence – for example, the fact that the bloodstain found on a suspect's sleeve is found to be a match for that of the victim – has been brought into even sharper focus. Statistics could potentially come to the rescue, but the complexity of the mathematics and the lack of agreement among statisticians make them a double-edged sword.

In statistics, Bayesian inference is a method used to estimate the probability of a hypothesis as additional evidence is learned. Bayes' Theorem links the degree of belief in a proposition before (prior odds) and after (posterior odds) the evidence has been taken into account. For example, suppose somebody proposes that a biased coin is twice as likely to land heads than tails. The degree of belief in this might initially be 50 per cent. The coin is then flipped a number of times to collect evidence. Belief may rise to 70 per cent if the evidence supports the proposition.

To add to the level of confusion, the expert witness and the jurors are actually considering opposite probabilities. What the forensic scientist is considering is, for example, the probability of a bloodstain found on the suspect's sleeve matching the blood type of the victim, given the proposition that the suspect is not involved in the crime. In other words, what are the chances of a purely coincidental match of blood DNA on the suspect's sleeve if the suspect had nothing to do with the victim or the crime? To put it yet another way: what is the probability of the DNA being a match if the suspect is innocent?

The jurors, on the other hand, are there exclusively to consider the probability of the suspect's innocence (and hence guilt) given the evidence. This might not seem so very different – but it is. The word 'given' introduces a conditional phrase, and confusing 'the evidence, *given* suspect's innocence' with 'suspect's innocence, *given* the evidence' is called 'transposing the conditional'. It happens so easily in court that it is more commonly known as the Prosecutor's Fallacy. Let's look at an example I used to give (sometimes successfully) when I was lecturing on the subject at Bramshill National Police Staff Training College

Let's take a game of 5-card poker played by you (the reader) for very high stakes, with one opponent. Your opponent is dealing and deals himself a straight flush (the winning hand). In Bayesian terms, the probability of dealing a straight flush is 72,193 to 1, *given* that the deal was fair. This is a mathematical fact based on the numbers and suits of a normal pack of playing cards. It is incontestable and does not vary – there are no prior or posterior odds to consider. It is equivalent to the scientific expert witness quoting the probability of the evidence (straight flush), given the innocence (fair deal) of the suspect.

This, however, is very different to the question of whether the dealer has dealt fairly, *given* he has dealt himself a straight flush. This depends very much on what you thought of the dealer before (prior odds) and after (posterior odds) he dealt himself the winning hand. It is equivalent to the juror considering the probability of innocence (fair deal), given the evidence (winning straight flush).

If your opponent is Harry Scrote (the well-known east London card sharp) you may well decide that the straight flush is sufficient evidence to call Harry a cheat. If your opponent is the Archbishop of Canterbury, you may well merely remark on His Grace's uncanny luck. The evidence is the same; the inference different.

So, what if you ask His Grace to play 'double or quits' and the same thing happens – another straight flush? The odds of two such hands in a row becomes billions to one against. Suspicious yet? A third repeat? Sooner or later, the odds against must mount sufficiently for you to question the card-playing credentials of even the Archbishop. Bayesian interpretation (as distinct from frequentist interpretation) expresses how a subjective degree of belief changes rationally to account for evidence.

The use of Bayes' Theorem by jurors is controversial. In the United Kingdom in 1996, a defence expert witness explained Bayes' Theorem to the jury in *R v Adams*. The jury convicted, but the case went to appeal on the basis that no means of accumulating evidence had been provided for jurors who did not wish to use Bayes' Theorem. The Court of Appeal upheld the conviction, but it also gave the opinion that 'To introduce Bayes' Theorem, or any similar method, into a criminal trial plunges the jury into inappropriate and unnecessary realms of theory and complexity, deflecting them from their proper task.'

★

It was around this same time that, back in the USA, the FBI entered the picture. Having seen the potential for the forensic use of DNA, the agency set up a research unit to establish DNA identification techniques for the Bureau. It was only after a full year of testing, ending in late 1988, that the FBI set up their own DNA laboratory at their Pennsylvania Avenue head-quarters.

Rather than being locked into a single technology or product, the FBI technique initially made use of a combination of four different DNA probes, including those developed by Lifecodes and Cellmark. The probes are themselves primers, and are the patented biochemicals used to identify individual genetic differences.

The main result of the FBI testing was that, finally, the standards that had already been established in the UK were emulated in the USA. The agency established laboratory protocols, performed validation studies and cut through the competing systems, methods and tools to establish a standard-ized system that is now in use in almost all labs in North America to this day.

By then, the Home Office-run HOFSS and Metropolitan Police Laboratory were taking over most of the primary forensic work on behalf of the prosecution, but other companies were working on the defence side of things. Cellmark had had a virtual monopoly of commercial DNA testing in the UK, but then, as the value of the market became clear, a competitor emerged in the shape of University Diagnostics, part-owned by University College London.

In September 1990 the then Metropolitan Police Commissioner, Sir Peter Imbert, first floated the idea of compelling suspected sex offenders to provide DNA samples to enable the police to build up a national index, or database, of genetic profiles at the International Police Exhibition and Conference.

Sir Peter said that DNA fingerprinting could help to solve a vast number of crimes of rape and other sexual offences if police were able to insist that suspects submitted to tests and if samples taken during previous inquiries could be offered as evidence in new cases. At the time, suspects could refuse a DNA analysis, and samples taken during investigations were destroyed once the investigation was complete.

Sir Peter accepted that some people might object to the proposal as an infringement of civil liberties. He also accepted that there was a need for much thought about how wide police powers to take samples should be, but he added, 'Whose liberties and whose freedom are we concerned about? The victim's or the offender's?' He suggested that the need for a public debate was urgent as, in the area of DNA analysis, science was threatening to overtake the law.

11

IT WAS A CASE THAT WAS OVER AND DONE WITH IN RECORD time.

Lunches at the Metropolitan Police Laboratory were usually staggered, to ensure there was never a time when any particular department was unstaffed, just in case an urgent case came in. On this particular Thursday afternoon I was the only one in the Sexual Assault Unit, and I was sitting at my bench completing some paperwork when a call came from reception.

A WPC had walked in carrying an item of clothing and was hoping that someone might be able to look at it immediately. I said I'd pop down and see what I could do.

When I got to reception, the officer explained that the item was a Babygro that might have been worn during a sexual assault. She explained that the child's mother had turned up at her local police station earlier that morning and claimed that her husband had been assaulting her baby.

The woman was understandably distraught and desperate to protect her child. She wanted her husband arrested and detained right away. Asked if she had any way of supporting such a serious allegation, the woman had produced the Babygro. She then said she had witnessed the sickening sight of her husband assaulting the child and that the abuse had ended with him ejaculating over the baby's clothing.

The husband had indeed been arrested and was being held at the police station. The police didn't have enough to charge him there and then, however, without having some analysis done on the item of clothing. If they had to wait too long for the results, they would have to let him go, and that could put both mother and baby at risk. Could I, the WPC asked with a hopeful smile, carry out a 'quick test' to see if there was semen on the Babygro?

It was a relatively quiet day and I didn't have any pressing casework at that time so I told the WPC to take a seat and that I'd do what I could straight away.

Back at my workbench, I first carried out a close visual inspection to see if I could identify any unusual staining. If I could, it would allow me to carry out tests on a specific area of the outfit, but if there was nothing visible to the naked eye, I'd have to test the entire surface. There were multiple stains all over the item but, considering it was a Babygro, that was to be expected. Nothing stood out as being particularly fresh, even when I used a magnifying lens.

At a crime scene, the next step would be to use an ultraviolet 'black' light to examine the outfit. Semen absorbs ultraviolet light and re-emits that energy as visible light. The same holds for other organic substances, and some bodily fluids – including urine – which will fluoresce when put under an ultraviolet light. Fortunately, seminal fluid usually fluoresces by far the most brightly.

However, this was in the laboratory, and the standard laboratory test for the presence of semen involves looking for acid phosphatase (AP), an enzyme secreted by the prostate gland that is present in large amounts in seminal fluid. The substance is not unique to the prostate and can be found in other biological fluids, including vaginal secretions, but the concentration is between 50 and 1,000 times higher in semen. Although a positive reaction to the test gives a strong indication that semen might be present, this then needs to be confirmed, usually by looking for spermatozoa under a high-power microscope.

I spread the Babygro on a sheet of plasticized paper, lightly sprayed it with water and then laid a large sheet of blotting paper on top. I marked the paper and the bench beneath it so that the paper could later be accurately realigned with the Babygro. After spraying the other side of the paper with water, I removed and cleaned the glass windows from two of the cupboards lining the laboratory walls and put them on the blotting paper. I added a few weights to press down on it gently. After a few minutes I removed the weights and glass, put the paper into a fume cupboard and sprayed the side that had been in contact with the Babygro with AP reagent. Almost immediately, a dark-purple stain appeared in the middle of the blotting paper. Although they were only presumptive, I'd done enough tests in the past to know that this reaction was more than vivid enough and it had happened so rapidly that it was fairly certain that semen was present.

Thanks to the marks I had made on the blotting paper, I could re-position it on the Babygro and mark an exact location for the suspected semen, which made it far easier to tell that particular stain apart from all the others. (See page 6 of the second picture section.)

The WPC had simply asked me to carry out a quick test to verify the wife's story, and the results up until that point seemed to have done this. However, knowing the consequences of a positive reaction for seminal fluid, I had to ensure the presence of spermatozoa by performing a confirmation test.

I cut out a small piece of the cloth from the area of the outfit where the 'suspect' stain was present and made a small extract. This I dried on to a microscope slide and stained with haematoxylin and eosin to enhance the contrast between the different parts of the cells present. I added a glass cover slip and took the slide to the high-power microscope. I brought the image into focus and could instantly see that some spermatozoa were indeed present. I had all the confirmation I needed, but, as I was looking through the microscope at the slide, I observed a large number of other cells as well.

I headed back down to reception and told the WPC, who had already been waiting for half an hour or so, that although semen was present, I had just spotted something unusual that I needed to account for and that I'd like to have the chance to investigate further.

The WPC agreed to wait, saying she would go and get some lunch, so I went back up to my bench and made up a few more microscope slides, using material from other parts of the stain. I wanted to be sure that there could be no mistake in what I had seen.

The more I looked, the more of the other cells I saw mixed among the spermatozoa. Although I was almost certain of what they were – they are quite distinctive if you know what you are looking for – their presence did not fit in with the story the WPC had been told.

Following my suspicions, I made yet more slides, this time using Lugol's reagent, which reacts with particular cell types, staining them brown, and my suspicions were confirmed. The other cells were from the wall of an adult vagina.

By the time the WPC returned, I was ready to report my findings. 'What you have here is a mixture of semen and vaginal material,' I told her. 'That's not what might occur with a man ejaculating on a baby. The only way this mixed staining could have got on to the clothes is if the semen were mixed with vaginal fluid first. It's vaginal drainage. The wife has had sex with her husband and then used this Babygro to wipe herself. '

The motivation was clear. The woman had wanted to get rid of her husband and had contrived this plan in order to do so. However, as is almost always the case, she simply hadn't known enough about the science to realize that her subterfuge could easily be spotted.

The husband was released from custody immediately, while the woman was charged with wasting police time.

DNA wasn't just changing the way that forensic scientists gathered their evidence and police investigated crime, it was also changing the way those accused of crime behaved.

In Canberra in late 1989 Desmond Applebee became the defendant in the first Australian court case to involve DNA evidence. After a blood test matched his DNA to semen found on the victim's clothes, he rapidly changed his defence from 'I wasn't there' to 'It was consensual.'

Calculating the amount of time that had passed since inter-course had taken place had long been of interest to the police and the forensic scientist in order to help support or refute the claims of the victim in rape allegations. Scientific papers had been written on the subject in the early 1960s. However, with DNA providing near-certainty of the identity of the source of any semen found, defendants were relying more frequently on the defence of consent.

This placed even greater emphasis on accurate assessment of when the offence had happened, based on the known time when personal body samples were recovered from the victim.

The defence regularly asked questions about how long semen persisted in the vagina or other orifices after intercourse.

The Metropolitan Police Laboratory had pioneered much of the research in this field, based on the persistence of spermatozoa in the vaginal tract. Indeed, one of the definitive papers on the subject had been written by my boss at that time. As usual in such matters, the research material was provided by the laboratory staff themselves – although I suspect they could not be described as 'long-suffering' in this case (unlike those giving control blood samples on a daily basis).

The research department provided a refrigerator in which female scientists and assistants would place internal vaginal swabs which they had taken themselves immediately after intercourse involving unprotected ejaculation. This intercourse was to have been effected in their own time, and not on the laboratory site, of course. Swabs were then to be self-taken at timed intervals, labelled as such, and supplied to the lab researchers. The 'less active' female members of staff were prevailed upon to provide negative control swabs. It was all a very delicate matter, requiring some tact and discretion on the part of the researchers, who clearly had a very up-to-date knowledge of their colleagues' sex lives.

While at the MPFSL, I spent several years, with the delightful support of three research students, Annette, Eileen and Sheryl, looking at the analysis of prostaglandin E (a lipid hormone found in high concentrations in seminal fluid) as a means of determining the time that had elapsed since intercourse, especially after azoospermic ejaculations – those containing no spermatozoa. The research was successful, but alas, no Nobel Prize – again.

One of my male colleagues took the quest for knowledge to an extreme and carried out a series of experiments on himself, injecting his own semen into his mouth and anus and then taking swabs at regular intervals in order to analyse and record the results. This was above and beyond the call of duty in itself, but then he was obliged to publish his paper in an international journal, in which he described, in forensic detail, just how the samples had been obtained.

Despite what you might see on TV, the amount of time that has passed between death and when a body is found can only ever be roughly estimated. Unless there is a witness or some independent record of the deceased's last moments, the time of death cannot be determined with any real certainty. The now-late pathologist Professor Taffy Cameron once told me, 'When it comes to time of death, the only thing you can guarantee is that it's going to fall somewhere between the time when the victim was last seen alive and when their body was found.'

I'd received training in the basics of forensic entomology – the use of insects to determine time of death – during my early days at the laboratory, but had not had a great deal of opportunity to use my skills until I came across the case of a dead body in a greenhouse.

There were too few cases around to get any real on-the-job training, so much of what I knew had come from a combination of reading books and identifying the remains of preserved flies and maggots kept in glass jars in the lab.

The first thing to learn is how to identify the developmental stages of this type of insect. A fly will develop from an egg through three 'instars' of larva (maggot) to form a hard-cased pupa before eventually hatching into an adult fly. Different flies take different lengths of time to reach their various larval stages, so a forensic scientist needs to know which species of fly they are dealing with in order to be able to identify the correct period of each larval stage. It's almost impossible for anyone other than a skilled professional entomologist to distinguish between two consecutive larval stages and, even then, they wouldn't be able to do so with the naked eye.

As is the case with so much of the 'forensic science' seen on television, the fiction is a very long way from the reality. In the opening scene of the very first episode of *CSI: Crime Scene Investigation* – a show that would quickly go on to become the most watched television drama in the world and forever skew public perceptions of the work of forensic scientists – lead investigator Gil Grissom pulls a maggot off a dead body in the bath and inspects it. 'Pupa stage three,' he declares. 'This guy's been dead seven days.'

Not a thermometer or temperature reading in sight, no calculations or microscopic examination of the larval stage, no suggestion that there may be a range of dates depending on circumstances – just 'seven days'. This is a very long way from reality.

What I would usually do is collect a few live maggots from the body of the deceased and grow them until they became adults. Once they turn into flies, they are much easier to identify. If the identification was straightforward, I would do it myself, but if it was too tricky I would take the flies down to the Natural History Museum and get one of their experts to help me out.

The maggots themselves tell only part of the story. In order to work out how long the victim has been dead, precise observations and measurements of the ambient temperature are needed. Flies won't settle and lay eggs until body fats begin to break down, and this is usually a function of how warm or cold it has been at the crime scene. So knowing this, when the body is found and in the days preceding, is crucial.

In this particular case, the calculation was made far more complicated by the fact that the body was in a greenhouse. For one thing, this meant that the ambient temperature was going to be higher than that outside, which would speed the process of decomposition; for another, the temperature could be subject to far more variation, as it could soar if direct sunlight suddenly came in through the glass panes.

As soon as I got to the scene I began collecting maggots. A few I killed by placing them in a tube of alcohol. I did this so that I would be able to compare their age at that point with those that I planned to take away alive.

The live maggots require a food source, unless you want to kill them slowly. I had been taught long ago that, if you want to keep your maggots happy, it's best to give them more of what they are already feasting on: the body.

If I was collecting maggots in a mortuary, I would ask the pathologist to slice off a part of a corpse's liver, as it is full of nutrients, but as this body was still fully clothed, it was easier simply to take a piece of the decomposing flesh surrounding the area in which the maggots were found and place it in a jar along with my maggots.

I'd been called out to the case late on a Friday night and there was no way I could get back to the office so, once I'd finished at the scene, I went home, taking everything with me. My girlfriend wasn't particularly happy about the lump of maggot-infested human flesh that remained in our fridge, next to the milk, for the rest of the weekend, but, as I explained to her, 'Maggots come with the job.'

I also needed to 'guesstimate' the temperature of the body around the time it was found. I borrowed a couple of recording thermometers from the lab, placing one of them inside and one outside the greenhouse so that I'd be able to come back after a few days and get a good idea of the fluctuations in temperature outside, along with the effect the greenhouse produces on the temperature inside. I also needed the local weather reports for the time leading up to the discovery of the body. Then it was back to the laboratory to undertake a series of calculations to work out what the ambient temperature in the greenhouse had been over the days before the body was found.

With all the data in place, I finally managed to estimate the time of death. And after all that work, just as Taffy Cameron had said, it turned out to be some time between when the victim had last been seen alive and when their body had been found.

Depending on how long it has been, there may be other ways of determining the time of death. Rigor mortis – the natural action that tightens and stiffens the joints – sets in around two to four hours after death and generally passes after around 24.

Identification of stomach contents helps to determine the type of food last eaten; however, estimation of the degree of digestion is also inaccurate in determining time of death because of the variability in how a person's system deals with different types of food. This can be greatly affected by the emotional state of the individual, too – digestion tends to cease if the person is frightened or fleeing. But it can be handy if you know the victim was last seen in a restaurant.

The analysis of stomach contents is gruesome work, and the smell is something I never got used to. There is some science involved – you can run tests to determine the presence of starch, tannins or alcohol, for example – but, in essence, it simply comes down to opinion. You put the stuff through sieves and then sit there, picking away: that looks like half a pea, that seems to be a piece of carrot, and so on. Tomato skins are at least easily identifiable.

A few weeks after the time-of-death case, I had a rather bizarre experience in the witness box at the Old Bailey that once again illustrates how the science and the significance of the scientific results can be poles apart. It involved a rape case in which I found semen which matched that of the suspect on one end of a tampon taken from the victim.

Apart from blood control samples, the tampon was the only item submitted to the laboratory; the results were unequivocal and I was hardly expecting to be called to give testimony. However, unknown to me, the suspect had admitted the intercourse but claimed it was with the consent of the victim. As I have already mentioned, it was often the case that, when the forensic evidence was good, the suspect changed his version of events from 'never even met the woman' to 'consent'.

I took the oath in the witness box and answered the prosecution counsel's obvious questions about finding semen on the tampon and the straightforward (pre-DNA) grouping results. It was then the turn of the defence.

One of the major tenets of cross-examination (I am reliably informed) is 'never ask a witness a question to which you do not already know their answer'. That way, counsel can try to ensure that statements contrary to their client's best interests are not aired in court. The defence barrister in this case had obviously never heard of this, or only thought – mistakenly – he knew what I was going to say.

'So Mr Silverman, I don't suppose you can help us as to which end of the tampon the semen was?'

'Actually, yes, I can. It was on the end with the blue string,' I offered helpfully.

Once again without my knowing about it, the counsels had been arguing that if the intercourse had been consensual then the victim would have removed the tampon before intercourse, replacing it afterwards; whereas if the tampon had been left in place during the act, then the supposition would be that it was rape. The significance of the scientific evidence had moved from the presence and identification of the semen to where on the tampon it was located. The judge (who really was wearing half-moon glasses) leaned forward and peered down at me over them from the bench.

'And what exactly is the significance of the aforementioned blue string?' he drawled in a very posh accent. And yes, he did say 'aforementioned blue string'.

We then had the bizarre situation of two male barristers, a male judge and a male expert witness all debating the purpose of the blue string. I was expected to come up with the answer – not really my field. But in those days tampons were frequently advertised on the television, and I had some knowledge of, if not personal experience in, their use. I (probably ill-advisedly) emulated the judge's questioning style.

'My Lord, I believe that the purpose of the blue string is to facilitate the removal of the said object.'

All hell broke loose. It was clear that my testimony indicated that the tampon was in place when the intercourse occurred – and hence supported the prosecutor's version of events – and it was the defence who had brought up the subject. We had a round of even more weird questions as to whether the 'said object' could have turned around inside the victim during intercourse. I claimed that the tampon was probably made precisely not to do that. The cross-examination ended with an irate defence barrister asking angrily, 'Mr Silverman – are you familiar with the female genitalia?'

A number of witty replies presented themselves to me, but contempt of court precluded me using any of them. Needless to say, the women on the jury were thoroughly enjoying the whole spectacle. Perhaps it would have been more sensible to ask one of them about the tampon.

12

BY THE TIME THE 1980s GAVE WAY TO THE 1990s THE WORLD OF forensic science was changing rapidly. DNA was now proving its value and was a mainstay of virtually every major investigation, and both the Metropolitan Police Laboratory and the Home Office Forensic Science Service (HOFSS) required a higher level of investment in order to take full advantage of the new technology.

As the procedure itself remained relatively expensive when compared to the cost of blood grouping tests, the fact that a growing number of items were being submitted meant the costs of forensic analyses were threatening to increase dramatically.

A key problem in both the Metropolitan Police and the HOFSS was that the lack of charging for the service led to a lack of accountability for the efficiency of the forensic science investigation. Not infrequently, I would write up a statement and send it off to the investigating team, only to be told the court case had already been heard or that the charges had been dropped.

On the other hand, some staff in the laboratory would spend huge amounts of time and devote a disproportionate amount of resources on particular cases, effectively turning them into research projects.

Back in 1989, the Home Affairs Select Committee report 'The Forensic Science Service' had highlighted the growing importance of forensic science but stressed that it needed to be properly managed and funded. The suggested solution was to introduce charging for the service. Consequently, in 1991 the Home Office Forensic Science Service Laboratories were restructured and awarded government executive agency status to form the Forensic Science Service (FSS).

Not everyone involved in making the decision agreed with it. An amendment tabled by Tony Worthington MP, himself a member of the Committee, urged his colleagues to reject the idea of charging police forces on a piecework basis, as he feared this would 'discourage police forces from using forensic science services on a speculative basis at the start of an investigation'. The amendment was defeated by two votes to one.

The Association of Chief Police Officers (ACPO) sent the committee a memo, stating that, in their view, charging would ultimately prove counterproductive. 'There are enough problems in the minds of investigating officers without adding the question: "Can I afford to send this to the laboratory?"'

Overnight, a cost had been put on the use of forensic science in the criminal justice system. The rocky foundations of a nascent forensic science market had been created. It was up to the new FSS Agency, and any other suppliers, to demonstrate the value of the service they were providing.

The changes would have little impact on the Metropolitan Police, which, having its own laboratory, was able to operate under its own rules, but, outside the capital, police forces began to prepare for the new regime.

Instead of having a capped amount deducted from their over-all budget each year to cover the cost of forensic science services, forces would each receive an annual sum initially ring-fenced for forensic science usage and have to manage this budget themselves. Forces around the country began reorganizing their forensic departments and appointing scientific support managers to control budgets. Previously, this was a role that had been carried out by relatively senior police officers, but, with the move towards a free market, it soon became clear to some forces that it was a role far better suited to someone with more specialist forensic science knowledge.

It was my boss at the Met Lab who first noticed that Kent Police were advertising for a scientific support manager and suggested I apply. I read the advert, trying to put any thoughts that my boss was trying to get rid of me to the back of my mind. In the end, I had to agree that it did seem to be a position to which I'd be well suited.

By that time, I'd been a reporting officer for 12 years. I'd worked on big cases and small cases, had my fair share of successes and made lots of good friends. I'd had good times and bad times in court and even managed to do a little foreign travel. I felt the time was right for a new challenge.

All the applicants who made it to the shortlist – around a dozen or so – were invited down to Kent for two full days of tests, assessments and interviews. The force insisted all candidates stay overnight and arranged a social gathering for the first evening to give all the department heads the chance to get to know the applicants in a less formal setting.

The first day was a bit like being back at school. I spent the morning sitting in a classroom with all the other candidates – many of whom I knew, and some of whom I worked with – and went through a series of exam papers to test our numeracy and literacy skills. After that we split into groups and, rather like something out of *The Apprentice*, were observed while we designed something by committee before being told to go off and get ready for the party.

The appointment was obviously being treated as a very big deal, as all the top brass of the Kent police were at the evening event. It seemed as though each of them had been given a 'target' candidate to get to know, apparently with the aim of asking one or two probing questions to make sure the candidates had answers to questions that may have remained unasked, unanswered or open on the application form. Typically 'politically incorrect' for the time, it seemed my inquisitor was determined to ascertain whether, as I was unmarried at 38, I was homosexual.

The next morning there was psychometric testing, which involved working through hundreds of short statements and questions that supposedly allowed the testers to determine your personality type, and the final stage of the process was the interview. And because this was effectively a brand-new position that was being created and Kent police didn't know exactly what they wanted from the candidate, everyone senior wanted to have some input into the decision. When I walked into the interview room I could hardly believe what I saw. There was a long table with six interviewers and, behind them, a further six. The board included the Chief Constable, the head of finance, the head of human resources, chief officers, head of fingerprints and scenes of crime and a selection of other hangers-on. And they all had questions for me.

I couldn't help but feel a little anxious. It occurred to me that I'd never had a proper job interview up to that point. I had walked into the position at the Metropolitan Police Laboratory because I'd worked there during my year in industry, and even the outcome of that student interview had been a shoo-in because of my tutor's connections.

Luckily for me, I was fairly relaxed about the whole thing. I answered the questions as best I could, telling them how I would tackle various scenarios, what my strategies would be for dealing with this and that, but it was clear that the interview was building up to the $64 million question: what experience did I have of running a budget?

The truth was, I had no experience whatsoever. The Metropolitan Police Laboratory did not charge for its services and had no intention of doing so at the time. If new equipment or staff were needed in order to meet demand, we obtained them. If too much work came in, we restricted the number of minor cases we dealt with. Charging didn't come into it at all.

I could have tried to bluff, but I decided to come clean and make a joke of it. 'I've had absolutely no previous experience, but I do have a gene for it,' I told them. The members of the board looked at one another, their foreheads wrinkled with confusion. 'What do you mean?'

'Well, I'm Jewish.'

It took a moment or two for the 'witticism' to work its way through all the members of the board, but it eventually did and it seemed to go down well.

A week later, I was offered the job. I then had to have a serious think about whether I wanted to take it or not. It would mean stepping away from bench work and no longer being involved hands-on in individual cases.

It would also be a management position. At the Met Lab I had a full-time assistant and usually one or two students working for me, but that was it. The job at Kent would put me in charge of 95 staff.

To add to the challenge, Kent wanted to civilianize their scenes of crimes department, which consisted at that time entirely of detective constables and sergeants, so all the experienced staff would be leaving and I'd be starting from scratch. The Metropolitan Police Laboratory was a slick, well-oiled machine, but at Kent everything would be new and I'd be going in at a time of massive change. Whereas with the Met, I could find the answer to most questions in a manual, in Kent that manual didn't yet exist. I'd be the one writing it.

The challenge was too exciting to resist. I accepted the job.

As I prepared to start my new post, the prospect of charging police forces on a job-by-job basis was beginning to create increasing disquiet. David Owen, president of the Association of Chief Police Officers (ACPO), predicted that the change would increase the cost of forensic services and pose big budgeting difficulties. Forces were likely to cut down on their use of forensic science services and to consider either setting up laboratories of their own or using private-sector contractors.

Even Janet Thompson, Director General of the FSS, had to agree that the introduction of charging was likely to hamper some criminal investigations, though she claimed they would almost always be minor offences. 'It is perfectly possible that in the past we did some work which was not very useful, such as spending £1,000 in the laboratory working on a £25 criminal-damage case. But it didn't happen very often. Because we now know precisely what our costs are and have an even better grasp of the police's needs, we will be offering much better value for money.'

In the hope of boosting its income, the FSS staged an open day in an attempt to secure its share of the market, promoting the fact that the services it offered could just as easily be used to identify the source of compounds causing river pollution or food contamination as to identify arsonists, forgers and burglars.

However, at around the same time, West Midlands police announced that they had been forced to devise stricter guidelines governing the use of forensic services because of the charging procedure. The force had allocated £1.25 million that year for forensic work, although it estimated that it would actually need £1.6 million.

The force decided to no longer send evidence to the Forensic Science Service from cases such as criminal damage of less than £250 or non-injury traffic accidents that did not involve drink-driving or other criminal offences.

The new guidelines angered detectives, who had come to expect that evidence gathered at crime scenes would automatically be forwarded to the FSS for analysis to provide the best possible case for prosecution. Under the new regime all evidence found at a crime scene would still be gathered, but a judgement about which items were to be sent for analysis would be made by scenes of crime supervisors.

In my role as scientific support manager for Kent, I had overall responsibility for photography, fingerprints and collecting evidence from crime scenes but I did not have a fully functional forensic lab. Instead, all forensic analysis was carried out by the Forensic Science Service in their regional laboratory in Aldermaston.

In order to get police forces used to the idea of being charged for its services, the FSS had decided to introduce the changes in stages. For the first year there would be 'soft' charging. Invoices would be issued so that forces could see how much they were potentially spending and budget accordingly, but no money would change hands. This was intended to make it easier to ascertain how large a budget each force would need – and easier for the FSS to ascertain how much income it could expect to receive and to make sure the books balanced.

The problem was, no one at the FSS really got their head round the fact that it was suddenly being required to function as a profitable business and the prices they planned to charge were not based on the actual cost of providing the services.

The initial pricing structure was worked out on the basis of the number of exhibits the FSS handled from each force. For example, it was worked out that the FSS needed to earn, say, £1 million from Kent, and knew it dealt with, perhaps, 10,000 exhibits each year from the force, so it calculated that if it charged £100 per exhibit, it would get the money it needed.

Things soon became farcical when arguments broke out about whether an exhibit was a single item or a bag of items. In particular, there was the issue of shoes. I would put 'a pair' of shoes in one bag but the FSS would argue that they had to be treated separately and would cost twice the standard fee. The argument then turned to the treatment of a 'pair' of trousers and soon fizzled out. After that, the FSS attempted to charge for its services by the hour. This seemed to make more sense, as this was how the staff doing the work were, effectively, paid and it was easier to work out how much was needed to cover costs and overheads.

The problem, however, was that not all the staff at the FSS worked at the same pace. Some reporting officers (ROs) would take days or weeks to carry out a procedure and charge accordingly, while others would get things done in a matter of hours, saving huge amounts of money. The customer was therefore, in some cases, pretty much being charged for the inefficiency of some scientists.

Once hourly charging came in, I started requesting that particular ROs carry out my casework, as I knew they would cost less. This was refused, so I then argued that the FSS needed to have an average charge rather than a specific one to take account of the different timescales involved in preparing different cases.

The other difficulty with the hourly charging was that some FSS staff were in the habit of turning a job into a major research project. At one point, I received a sample of white powder from a PC who had been out on patrol and come across a youth who was firing an air gun using pellets containing this white powder, which 'exploded' on impact.

I sent to the laboratory a simple request: was the powder an explosive substance? What we got back was a lengthy report showing that the powder had been subjected to every single test available, including a very clever inorganic synthesis of a fresh sample to show how it was made. There was a full breakdown of its chemical components, and all sorts of data about its possible origin. It was a hugely impressive piece of research, but the bill that came with it was for £32,000.

In response to this and similar incidents, Bob Green – one of my senior SOCOs – put together a comprehensive flow chart which we then enclosed with every submission to the lab. It was a complete breakdown of exactly which tests to carry out, which order to do them in and when to call myself or one of the other staff at Kent in order to get permission to carry out further testing.

Naturally, the reporting officers at the FSS hated the flow charts and, soon after they had been introduced, I got a visit from Gary Pugh, then an account manager for the FSS at Aldermaston, who invited me out to the laboratory to speak to the staff there.

I expected a confrontational meeting with them – so far as they were concerned, we were telling them how to do their jobs and doing it in an incredibly patronizing way. I explained that I had a budget to manage and simply wasn't able to allow them to do whatever they wanted the way we all had in the days before charging. I also explained my own philosophy when it came to forensic science: what is the point of carrying out a particular test if you don't know any more after it than you did beforehand?

However, I could not help but think back to the time the police officer had brought the Babygro into the Metropolitan Police Laboratory and asked me to take a look at it. With no restrictions, I was able to take the extra time necessary to check for the presence of vaginal cells and prevent an innocent man being held in prison on remand until his trial. If that had happened after charging had been introduced, the instructions supplied with the exhibit would have been to test for the presence of semen and nothing else, and the end result might have been very different.

Eventually, I came to an amicable agreement with the FSS. We agreed to stop sending the flow charts and they agreed not to carry out additional, unwanted research on our submissions.

When I started at Kent Constabulary in 1990, all scenes of crime officers working anywhere in the country (with the exception of the Metropolitan Police) were being trained at the national SOCO training school at Durham. As Kent wanted to replace all its SOCOs with civilians, I needed a large number taught at once, but Durham could only offer five places each year, so I set up my own training school, putting together a curriculum 'borrowed' from Durham with all the things I felt the recruits needed to know.

With so many police SOCOs to replace with civilians, it was not possible to recruit trained SOCOs from other forces. We would be obliged to take people from other professions with no experience at all and, in the space of six weeks, feel confident to send them out to burglary scenes to gather all the necessary information. The course I instigated still runs, and Kent is the only force, other than the Met, that trains its own SOCOs. We were very lucky to have no murder in Kent for six months. A never-to-be-repeated run of peace!

I helped out with training SOCOs in blood pattern analysis

and forensic science awareness. We made multipurpose crime scenes by creating blood splashes and distributions on large wallpapered boards to represent kicking, battering, punching, etc. These slotted seamlessly on to the walls in the 'crime scene house' – an unused police house at HQ which I had commandeered. Using a variety of devices, including a spring mousetrap to which I had affixed plastic plates, I threw horse blood around the walls in close facsimile of the various blood patterns the SOCOs might encounter. I primed the plastic base plate of the mousetrap with blood and pulled back the spring with the upper plate attached. As the spring snapped shut, the two plates hit each other, projecting the blood in a very good approximation of a medium-velocity impact such as one might encounter in a battering with a cricket bat. SOCOs were let loose in the house, set up in various scenarios and filmed on CCTV going about their examinations. It was innovative at the time, and was later copied by the national training school.

In-house training saved money, and I saved even more by making my own KM reagent. The FSS was charging something like £10 for a small bottle, and it went off quickly and we used a lot of it. By buying my own still, I was able to make enough to last the force for years, at a cost of around £50. Charging a high price for reagents that had only a limited shelf life was one of the ways the FSS hoped to boost its income but, ultimately, it backfired, in Kent at least.

Another, more insidious, ploy by the FSS was to try to drive a wedge between the Police and the Crown Prosecution Service. One morning I had a phone call from our regional Chief Crown Prosecutor, explaining that a few days earlier a representative from the FSS had been to see him regarding the new charging arrangements. He was surprised at the request for the meeting, because the CPS wasn't involved in paying for forensic science services – that was the responsibility of whichever force was investigating a case. During the meeting, the FSS representative suggested that the police would jeopardize the prosecution of cases by failing to send all evidence to the laboratory as they had done in the past, in the days before charging. The FSS suggested that the Crown Prosecutor introduce a policy whereby all evidence was submitted to the laboratory and that the decision about what to examine – and therefore how much to charge – would be made by the scientists themselves. It was an attempt to boost profits and bypass the gatekeeping role of budget holders such as myself but, thankfully, the CPS refused to play ball.

One of the unintended consequences of charging was the eventual creation of an ACPO National Firearms Database. Before, firearms which might be found when searching a suspect's home or which had been discarded in the street would be sent to the FSS to be compared with an existing firearms database. This contained the data from all UK offences where a firearm had been discharged and bullets or cartridges found. The intention – often successful – was to match the weapon with one used in a previous crime.

The problem with charging was that the police force finding the weapon would pay for the submission to the FSS laboratory. The case that was cleared up as a result was, however, all too often in another police force.

Submissions of firearms for speculative searches dried up. The intelligence was vital, so ACPO took it upon itself to organize, manage and centrally fund a National Firearms Database.

Although I was principally employed in a managerial role, the powers that be within the Kent police were not blind to the fact that, for the first time, they had a fully trained forensic scientist on the staff. Whenever certain cases came in – most commonly those involving some element of blood spatter – I would be asked to lend a hand. I was always happy to oblige, and it helped make the sudden transition from proactive reporting officer to desk-bound supervisor a little easier.

One of the first cases I was called out to involved a fatal hit-and-run accident in which the victim was a young boy. By the time I arrived at the scene the body had already been taken away and only the diminutive chalk outline remained.

It didn't take long for me to ascertain that the multiple patterns of blood in the tracks leading to and from the site of the body demonstrated that, while the victim might well have been struck accidentally the first time, the car had then reversed back over the body at least once in what could only be assumed to be an effort to ensure the boy was dead.

I informed my colleagues: they were not dealing with a hit-and-run after all but with a murder.

Many of the incidents I dealt with at Kent were, as they had been in London, disturbing and upsetting, but this wasn't always the case.

I was sitting at my desk at Police HQ in Maidstone one sunny Friday after lunch. The phone rang and a senior investigator, who I knew well, said that he wanted a forensic pathologist at a murder scene near to HQ. I had control of the budget for the expensive resource of pathology and so wanted to know the circumstances to approve the call-out of a pathologist – well over £1,000 at that time.

I asked for more details and was told that some builders had been doing work in the garden of a private house and two of them had been out to the pub for lunch. When they returned they were told to dig up a tree root, which they started to do while their colleagues went for their lunch – whereupon they unearthed a human skull! Then called the police.

The senior investigator was adamant that this was a recent murder, as there were maggots in the eye sockets.

'Are you sure?' I inquired. 'Maggots? Actually wriggling maggots in the fleshless eye sockets?'

'Yes! There isn't a scrap of flesh left.' He was getting angry with me.

'And the workmen unearthed the skull and maggots when their workmates went to lunch?'

'Look. Can I have a pathologist out urgently?'

'Not a chance,' I replied, 'but I'm coming over myself.'

I got the address and was there in ten minutes.

My reception was cool to say the least. The senior investigator clearly thought I was impeding his murder inquiry and would report me as having done so.

I asked to see and was shown the skull, just as it was unearthed, with the fork still stuck in the ground next to it. Sure enough, it was shaped as only a human skull is shaped, and white and clean and shiny. Not a hint of flesh, but both eye sockets alive with writhing maggots and a kind of coarse brown powder.

I knelt down to get a better look and prodded the maggots with my finger.

'Is there a fishing shop near here?' I asked the investigator.

'Just around the corner,' replied his detective sergeant. 'So what?'

The second two workmen were back from lunch now and were standing off to one side watching the procedures. I knew by then what to look for, and I could see them smirk.

'Come here and I'll show you.'

I explained to the senior investigator that maggots require sustenance and there was clearly no flesh on the skull. Furthermore, the empty eye sockets would never have contained the putrefying fat which would attract flies. Also, the maggots would have been adult by now and flown away, and maggots don't live underground, where the skull was discovered. He was unmoved.

I pointed out the brown substance with the maggots.

'I'm not an angler, but I'd bet that is sawdust. These maggots are live fishing bait.'

He was wavering but, clearly not wanting to lose face, he stuck to his request for a pathologist.

'For pity's sake!' I was exasperated now. 'Look at this!'

I had pulled the skull out from its resting place and turned it over. The words 'Made in China' were clearly visible on the bottom.

Having a sense-of-humour failure, the senior investigator charged the two workmen with wasting police time.

13

SOONER OR LATER, EVERY TV DETECTIVE DRAMA SHOWS A fingerprint being fed into a computer which, a few moments later, flashes up the word 'match' on its display screen, often accompanied by a picture of the suspect. Despite the huge processing power of modern computers, however – they are able to search through up to 500,000 prints in a single second – fingerprint computers never actually provide investigators with a single suspect. Instead, the system produces a list of potential matches in order of closeness, and these are then examined manually by a human operator or fingerprint expert. Before they are considered final, these matches are verified by other qualified fingerprint experts.

Although everyone's fingerprints are, theoretically, unique, there are prints which can be very similar and this can occasionally cause difficulties.

Following a series of terrorist bomb attacks on commuter trains in Madrid, Spain, in March 2004, which left nearly 200 people dead and injured more than 2,000, Spanish police investigators recovered several poor-quality, partial fingerprints from detonator caps found at the scene of one of the blasts.

These prints were submitted to the FBI for analysis and fed into AFIS – the automated fingerprint identification system. The system generated a list of around 20 possible prints, one of which was then triple-checked by hand and subsequently declared to be a 100 per cent match to Brandon Mayfield, a lawyer from Oregon.

Mayfield seemed to tick many of the boxes in the FBI terrorist profile. A former officer in the US Army, he had married an Egyptian national and converted to Islam. A year earlier he had been working pro bono on the case of one of seven men accused of travelling to Afghanistan to fight on the side of the Taliban.

Mayfield was arrested and held for two weeks while further investigations were carried out. When the Spanish police did their own work of comparing Mayfield's print with the one re-covered from the scene, however, they found that, while there were eight points of similarity between them (only half of the 16 'required' – there was no official standard – for a match in the UK), they were far from a 100 per cent match.

The problems were that the initial print was of poor quality, that it provided only a partial view of the entire finger and, perhaps most crucially, that the FBI experts doing the manual check had been told in advance that Mayfield seemed to be a strong suspect, which had possibly affected what should have been an impartial fingerprint examination. Mayfield was re-leased without charge. He later received an apology from the US government and a settlement of $2 million.

It was a similar story when 51-year-old Marion Ross was found murdered in her Kilmarnock home in January 1997. Her ribs had been crushed and she had been stabbed in the eye and throat.

She was known to be security-conscious and there were no signs of forced entry at her home, so detectives from Strathclyde Police were convinced she either knew her killer or had willingly let the person responsible into her home.

A fingerprint found on a gift tag on an unopened Christmas present was linked to David Asbury, a handyman who had once worked at the Ross home. A print on the bathroom door was then positively identified by four experts – with a total of 100 years' experience between them – as belonging to Shirley McKie, a detective constable with Strathclyde Police.

McKie denied ever having been in the house, but was suspended. Asbury was put on trial after a print belonging to Ross was found on a tin box in his home. McKie was then called in to give evidence by his defence team, who realized that, if they had made a mistake about her print, it was possible they had made a mistake about Asbury's as well.

Asbury was found guilty and McKie was charged with and convicted of perjury. At her appeal, American finger-print expert Pat Wertheim spent six hours in the witness box, demonstrating that his fellow experts had made an error and that the print did not belong to McKie after all. Instead of the 'required' 16 points of similarity, he could find only five.

'As soon as I looked at the print and realized it was the wrong identification my stomach knotted up. It was a terrible feeling,' he said. 'The danger in the system is that if one person makes a mistake, then they give it to a second person and say, "This is Shirley McKie's thumbprint," the second person automatically has a mindset expecting it to be that print. That type of verification is prejudiced before it begins.'

McKie's conviction was overturned and she was awarded £750,000 compensation. Asbury's conviction was overturned when the Court of Appeal accepted the fingerprint evidence in his case was unreliable.

What both cases clearly demonstrate is that, despite the way fingerprint evidence is portrayed in the media, all comparisons ultimately involve some human element and, as a result, they are vulnerable to human error.

As increased numbers of DNA analyses confirm that no two individuals' profiles will be a complete match, some have called for a similar level of rigour to be applied to fingerprint comparisons.

In a speech to the Forensic Science Society, Lord Justice Leveson, an appeal judge and chairman of the Sentencing Council, called into question the whole notion of fingerprints being infallible, citing 'numerous' cases in which innocent people had been wrongly singled out by means of fingerprint evidence.

No two fingerprints are ever exactly alike in every detail – even two impressions recorded immediately after each other from the same finger. It requires an expert examiner to determine whether a print taken from a crime scene and one taken from a subject are likely to have originated from the same finger.

Unlike other forensic fields, such as DNA analysis, which give a statistical probability of a match, fingerprint examiners traditionally testify that the evidence constitutes either a 100 per cent certain match or a 100 per cent exclusion. However, one study found that experts do not always make the same judgement on whether a print matches a mark at a crime scene when presented with the same evidence twice.

Six examiners in several countries were given eight sets of prints to compare on two different occasions, without knowing it was part of a study. They changed their decision in six cases, and only two of the experts were consistent with their previous decision.

The research, carried out at Southampton University,

corroborated Wertheim's conclusion in the McKie case: that examiners were more likely to change their decision if given contextual information (such as 'the suspect has confessed') which conflicted with their previous judgement.

Prior to the introduction of fingerprint analysis, one popular method of criminal identification in England during the late 1800s was the Tattoo Index. Since it was noted that many in the criminal profession were 'addicted to tattoos', it was deemed advisable to begin detailing their location and appearance. Initially, dozens of identifications were made through this index but, quite predictably, criminals soon developed the habit of periodically altering their tattoos.

At that time, police and prison authorities around the world were looking for ways to deal with an ongoing problem: there was simply no accurate way to identify, and thereby appropriately incarcerate, recidivists. Too many hardened criminals were being sentenced as first-time offenders.

The solution seemed to be Bertillonage, a system invented by French ethnologist Alphonse Bertillon, who took measurements of certain bony portions of the body, among them skull width, and length of foot, forearm, trunk and left middle finger. These measurements, along with hair colour, eye colour and front- and side-view photographs, were recorded on cardboard forms. By dividing each of the measurements into small, medium and large groupings, Bertillon could place the dimensions of any single person into one of 243 distinct categories. Further subdivision by eye and hair colour provided for 1,701 separate groupings.

The system was soon adopted around the world but, in 1893, a fatal flaw was discovered by prison warden R. W. McClaughry when he booked a man named Will West into Levenworth Penitentiary in Kansas. McClaughry was convinced he had seen the man before, though West denied having any previous convictions.

Looking through the Bertillonage records, McClaughry found a card with the exact measurements he had just taken for a prisoner named William West. The picture looked exactly like the man in front of him. The only problem was that William West was already in prison, having been sentenced more than two years earlier.

Despite sharing a name and looks, the two men were not related; they just happened to share Bertillonage measurements. McClaughry was soon on the lookout for a better identification system and, while at the World Fair in St Louis in 1904, he met Sergeant John K. Ferrier from Scotland Yard, who was there to guard the Crown jewels.

The pair got talking and Ferrier explained how Scotland Yard had been using fingerprints for the past three years. When McClaughry checked the fingerprints of Will and William West, they were completely different. Even identical twins, who will produce identical DNA profiles, have different fingerprints.

The use of fingerprints had begun in 1858 with Sir William James Herschel, Chief Magistrate of the Hooghly district in Jungipoor, India. Unable to identify themselves with a written signature, Herschel asked locals to stamp their business contracts with their palms. He did this on a hunch that it would be a good way of identifying someone, not because he knew the science behind it.

In the 1870s Dr Henry Faulds, the British surgeon-superintendent of Tsukiji Hospital in Tokyo, Japan, took up the study of 'skin–furrows' after noticing finger marks on specimens of 'prehistoric' pottery. A learned and industrious man, Dr Faulds not only recognized the importance of fingerprints as a means of identification, but devised a method of classification as well.

In 1880 he forwarded an explanation of his classification system and a sample of the forms he had designed for recording inked impressions to none other than Sir Charles Darwin. Darwin, in advanced age and ill health, informed Dr Faulds that he could be of no assistance to him, but promised to pass the materials on to his cousin, Francis Galton.

Galton's primary interest in fingerprints was as an aid in determining heredity and racial background – in part the same as Alec Jeffrey's initial interest in DNA a century later. While Galton soon discovered that fingerprints offered no firm clues to an individual's intelligence or genetic history, he was able to prove scientifically what Herschel and Faulds already suspected: fingerprints do not change over the course of an individual's lifetime, and no two fingerprints are exactly the same. According to his calculations, the odds of two individual fingerprints being identical were one in 64 billion. Galton went on to design a form for recording inked fingerprint impressions and defined three main pattern types: loops, whorls and arches.

At this time, in British India, Edward Henry, the new administrator of the Bengal district, was experiencing the same problems with the local population as Herschel had.

Henry was convinced that a system of identification based solely on fingerprints was possible. A correspondence, and subsequent friendship, was begun with Galton, and in 1894, when he went to England on leave, Henry visited Galton at his laboratory to learn more and iron out the kinks in his classification system.

After returning to India, Henry had fingerprints and Bertillon measurements taken of all prisoners under his jurisdiction. In a report dated 24 June 1896 he outlined the advantages of a fingerprint-based system over Bertillonage, but admitted that such a system had not yet been worked out.

This classification system, allowing for 1,024 primary groupings, was instituted in Bengal in early 1897. The success of Henry's experiment encouraged him to make a formal request to the government of India for the appointment of an independent committee to review his new system and compare it with Bertillonage.

The committee met in March of 1897 and submitted a report to the government which concluded that the fingerprint method was far superior to the Bertillon system. On 12 June the Governor General signed a resolution directing that identification of criminals by fingerprint impressions was to be adopted in British India.

Henry's Fingerprint Classification System was so successful there that a second committee was assembled to review Scotland Yard's identification practices. This committee in 1900 recommended the total abandonment of Bertillon's anthropometric system in favour of the new Henry System, a variation of which remains in use to this day.

There are two main types of fingerprint – patent and latent. The first type occurs when the skin of a finger contaminated with something makes contact with the surface of an object, leaving a 'ridge impression' that can be seen with the naked eye. A sub-group (sometimes referred to as 'plastic prints') are also visible to the naked eye and occur when a finger leaves an indentation in a soft material such as clay. The latent print can normally only be seen with enhancement – a special powder, for instance. Latent prints come about when the finger secretes sweat on an object, leaving an invisible pattern on it.

Every finger has a different pattern of ridges which investigators analyse and, so far, no two fingerprints among the billions recorded have ever been found to be the same. In February 1939 the *News of the World* offered the then enormous sum of £1,000 to any reader who had a fingerprint identical to the one printed on its front page. No one was able to claim the money.

When I arrived at Kent, fingerprints were still being held on a series of cardboard index cards and were searched by hand. Although it was slow, the system was relatively efficient, as the quality of matches it provided was generally very good. The staff in the fingerprint bureau developed a knack for the work and many of them were said to be able to recognize and name certain prolific criminals simply by seeing one of their prints. Having done the work for so many years, the unique features of some prints had become embedded in their minds.

There remained, however, a desire to upgrade the system and, as fingerprinting came under my department, I was told that some money had been put aside for the acquisition of a computer. At that time, systems were few and far between. One was being developed in the UK, but was likely to be some years away, and there were two other main competitors – MORPHO in France and PrintTrack in the USA. Accompanied by the head of my fingerprint bureau, I set off on my travels with the intention of trying both systems out.

I had a theory – if the machine was doing what it was supposed to do, then anyone with a bit of training should be able to recover a print. My colleague Mike, an experienced fingerprint expert, held the opposite view, believing that only someone who was already a fingerprint expert would be able to get the machine to search properly in order to generate the correct hit.

There was only one way to test the theory out – with a little competition. First in Paris, and later in Los Angeles, we were shown how the machines worked and invited the operators to feed in a sample print which we had brought with us. The machines looked at prints in different ways but in both cases the operator had, in one instance, to identify which parts of the unknown print they wanted the system to look for and, in another, simply to leave it to the machine to decide what points were of greatest significance.

For the contest between myself and the other Mike, I was the one to let the machine do everything on its own, while he used his knowledge to adjust the settings. We both managed to get the right print as part of the selection but, every single time, his was far higher up the list of possible matches than mine was. It was rock-solid proof that, even when the system was fully computerized, skilled human operators would still be needed.

I knew that the staff at the bureau and particularly those in the fingerprint department of Scotland Yard would be extremely relieved by this. They had all feared the loss of their jobs once computers came in. In a bid to ensure this would not happen, they had initially lobbied to take over DNA work. The problem was that Alec Jeffreys had originally called his discovery 'DNA fingerprinting', leading the fingerprint bureau to believe that the bar-code patterns resulting from DNA analysis could be read in the same way as fingerprints and should therefore come under their remit.

We tried to explain to them that the process was far more scientific and required a completely different set of skills and knowledge, but they were adamant. I think one of the reasons the name of the technique was eventually changed to DNA profiling was to prevent there being any confusion, though it did also reflect the switch from MLPs to SLPs. The former, though difficult to read and time-consuming to produce,

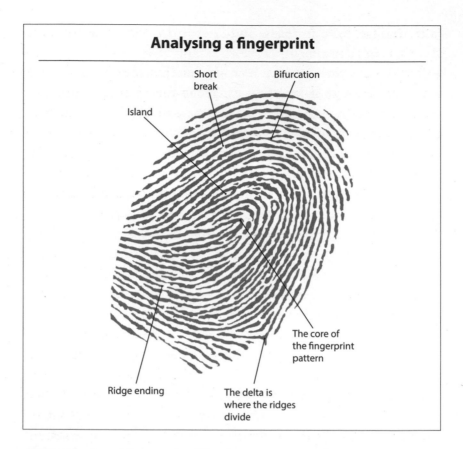

Analysing a fingerprint

Short break

Bifurcation

Island

The core of the fingerprint pattern

Ridge ending

The delta is where the ridges divide

did indeed produce a profile that was every bit as unique as a fingerprint, while the chances of individual SLP profiles being shared was far higher.

When identifying prints, the Automated Fingerprint Identification System (AFIS) computer uses two main features – ridge endings and bifurcations. A ridge ending is exactly what it sounds like. Some fingerprint ridges make continuous patterns, especially those that form whorls and arches, while others stop suddenly. A bifurcation is when a ridge splits into two. Other features include ridge dots. Also known as islands, these are ridges that are only as long as they are wide. Computers can also look for cores (the central circle of a whorl) and deltas (a triangular pattern created where three other patterns come together on a finger).

In essence, the computer looks for these points and then compares them with the five nearest similar points. Measurements are made of distance and angles, and all this information is fed into an algorithm to produce a digital number with a particular value. Prints with a similar value should look similar, but this is not always the case. The way the calculations are put together means that similar values can be produced by prints that look quite different, hence the list of possible matches and the need for a final check by a human operator. The machine doesn't find the print, but it makes the whole process of searching much, much faster.

We first had to devise new procedures for backing up the collection, in case of a catastrophic failure. We knew that we needed the system for only two or three years, because after that the national system would be introduced and Kent would be able to share it.

The national system is far more useful for major crimes. When it comes to burglaries, most of the offences are carried out by local criminals, so there is little benefit in sharing a system with an investigating team miles away for that kind of crime. However, buying the system was one thing. Back-converting all the existing Kent records so they could be entered on to the database was quite another.

In addition to the classic ten-print forms, we also used to keep a collection of palms and palm edges. We found that many burglars, when checking to see if a house was empty or what goods there were inside worth stealing, would put the edge of their palm up against the window to act as a shade to enable them to see more clearly. Because of this, we had a collection of palm-edge prints that contained similar patterns to fingerprints themselves and are equally unique. It was not possible, however, to load these into the machine, as the algorithm would not have been compatible.

With only one copy of each set of fingerprints, we didn't want them to go out of the country in order to be fed into the computer database. We also needed the prints in order to carry on with our day-to-day investigations. In the end, the solution was to hire a series of industrial photocopiers and duplicate the entire fingerprint collection.

Even the copies were classified, so they could not be sent over to the USA without being accompanied. Officers from the fingerprint bureau, working on rotation, spent weeks in California so they could make sure the copies were safe and that the prints were being entered into the computer in the most efficient fashion possible. There was no shortage of volunteers to baby-sit the copies. Nice work if you can get it!

14

FOR A WHILE IT SEEMED AS THOUGH THE WHOLE WORLD HAD
turned upside down.

I had made my way down to a hotel on the cliff front in
Folkestone on the Kent coast in order to examine some blood
spattering at a murder scene. The derelict Hotel de France had
been the subject of an arson attack and had been used by some
in the area who were living rough. There were various blood
distributions around the blackened room, but the main one
was on the edge of a small sink attached to the far wall.

The first time I saw it, it seemed to make little sense. The
only way the patterns could have been created would have been
if the victim had been pinned to the ceiling and then allowed
to fall down vertically in such a way that his forehead collided
with the edge of the sink.

It took some time to work out what had happened. Although
the sink appeared to be attached to the wall, it wasn't, and the
lead drainpipe to which it was fixed was flexible enough to
allow the entire sink to be moved down to ground level. The
killer had knocked his victim to the floor and then brought

the edge of the sink down repeatedly on his head while he was there, before returning the basin to its upright position. The distribution of the blood spatter clearly indicated that the sink had been used as a weapon but, even more interestingly, there was a clear fingerprint in blood on the edge. The finding of the suspect's fingerprint, in the victim's blood, on the murder weapon, is the Holy Grail of scientific investigation of murder, so I took the sink back to my fingerprint department at HQ in Maidstone.

The upshot of the examination was that the fingerprint matched one of the vagrants using the hotel and it was, indeed, in the blood of the victim. I wrote a statement of my part in the highly successful examination – only to be firmly rebuked by the detective superintendent in the case. Apparently, I had taken too long to get a result (although it was in fact months sooner than it would have been had I called in the FSS). There's no pleasing some people!

By coincidence, the evening of the day that I visited the Hotel de France, I had been invited to give the inaugural civilian after-dinner speech at the Kent Constabulary officers' mess.

Civilianization of senior police posts was, at that time, quite a rare occurrence and Kent was trying hard to make us fit in. It had been decided that civilians of a grade equating to super-intendent would be eligible to join the mess. It was considered an insult not to take up the offer. Quasi-military clubs are not really to my taste, but there were three civilians of appropriate seniority and it was the done thing to join. Not all the other members of the mess were happy with the interlopers, however.

To my horror, after just two attendances I was approached by the Deputy Chief Constable and invited to give the after-dinner speech in honour of the guests – a first for a civilian. It was the equivalent of a best man's speech at a wedding and I was expected to be witty and entertaining. I politely declined but was firmly told that the Chief Constable, Paul Condon, expected my acceptance – or my resignation. At that point I graciously accepted. I knew that my performance would be closely scrutinized.

Having agonized for weeks over this speech, and expecting to polish my presentation on the day of the mess dinner, I was instead rushing back from Folkestone – dirty, late and having had no chance even to read through my speech notes. All I had was some brief notes on the guests.

I managed a shower, but the bow-tie proved too much for me and I reverted to a pre-tied one (definitely *not* the done thing). In the event, the lack of preparation turned out to be my good fortune. Unable to rehearse into a dull recitation, I had to wing it and improvise a lot. One guest was the brother of our finance director, Mr James. He was an accountant, and both he and his brother had international sporting medals in rifle shooting. I didn't know either man well, or anything about the chief's sense of humour, but on impulse I started comparing the FD and his brother to 'those other James boys, out in the Wild West, who also had a liking for guns and other people's money'.

The other guest was Dr Mike Heath, one of the forensic pathologists we employed in Kent. I had known Mike from my previous career at the MPFSL and so gave him a big build-up. This was something which was to return to haunt me much later.

The rest of the evening passed in a blur, but I do clearly remember being congratulated for an exemplary performance by the chief. I had got away with it.

★

While I had been settling into my new role at Kent, Robert Napper, the rapist whose samples had been destroyed as the result of a mix-up at the Met Lab in the summer of 1989, had got over his remorse and returned to his old ways.

On 10 March 1992 he attempted to rape a woman after attacking her from behind in an alleyway. Although he had a knife, the woman managed to escape, though Napper did get so far as to ejaculate on her clothing, making it possible for a DNA sample to be obtained.

A week later, Napper was responsible for another attempted rape, this time on open ground in King John's Walk in Eltham. When the woman tried to resist, Napper made good on his threats, stabbing her in the breast. He ran off but, once again, left behind a semen sample on the woman's clothes.

Although it was far better to have DNA samples than not, they were of little help at that time. With no suspect and no computerized database of individuals' DNA to compare the samples against, having them did nothing to take the police any closer to putting an end to this spate of violence.

In May 1992 yet another woman was attacked, as she took a walk with her two-year-old daughter in a pushchair on King John's Walk. She was subjected to a frenzy of violence, during which she was punched several times. Once more, a DNA profile was obtained but, yet again, it did nothing to move the case forward – other than to have conclusively linked the three attacks as having been committed by the same offender.

In June 1992 police set up Operation Eccleston to investigate the attacks, all of which had taken place within a short distance of a network of pathways known as the Green Chain Walks, as well as a number of other incidents that were also believed to be linked.

Police thought they were looking for a violent, sadistic rapist – a man with no qualms about using violence in order to terrify his victims into complying with his instructions. Although each and every one of the Green Chain rapes had been brutal and uncompromising, police did not think they were looking for a killer. They could not have been more wrong.

On the morning of 15 July, a month after Operation Eccleston had been established, Napper moved into a whole new league. As she walked across Wimbledon Common with her two-year-old son, Alex, and the family dog, Molly, former model, lifeguard and would-be children's television presenter Rachel Nickell was subjected to a horrific attack in which she was sexually assaulted and stabbed a total of 49 times.

The only witness was her son Alex, who was found clinging to his mother's blood-soaked body, desperately trying to wake her up. Police spoke to everyone they could find who had visited the common that day and several gave a description of a white man carrying a black bag being there at the time of the attack. He was later seen washing his hands in a stream and was noted to have been wearing a belt on the outside of a white top. Alex seemed to confirm that this was the man responsible, telling the police about a 'man with a black bag who attacked Mummy'.

Tapings were taken from Rachel's body at the scene to establish if the attacker had left any fibres on her during the assault. A cast of a footprint believed to have been made by the attacker was also taken, and flecks of red paint were recovered from Alex's hair. Although substances believed to be body fluids were found, there was not a sufficient quantity for DNA tests to be carried out.

By the end of August, police working on the rape investigations as part of Operation Eccleston had put together an e-fit of the suspect, compiled from descriptions provided by the victims and, soon after it was released, a member of the public made contact and suggested that it bore a striking resemblance to Robert Napper. Despite there being nothing to link him to the murder of Rachel Nickell, it seemed his reign of terror was at last about to come to an end.

Police visited Napper's home and asked him to attend Eltham police station on 2 September to provide a blood sample for DNA purposes. As part of their report on the meeting, officers recorded his height as being either six foot one or six foot two.

Napper failed to turn up for his appointment at Eltham station but, the next day, a call to *Crimestoppers* (the television programme used by the police to ask for the help of the public) once again named him as a likely candidate for the Green Chain rapes.

Police returned to Napper's home, but he was not there, so they left a note asking him again to attend Eltham station, this time on 8 September, to provide a DNA sample. Once again, he failed to keep the appointment.

Meanwhile, detectives investigating the murder of Rachel Nickell had little in the way of forensic evidence to go on and decided to try a different approach. Criminal psychologist Paul Britton was called in and asked to create a profile of the offender. At the same time, police questioned around 32 dog-owning men who regularly walked them on the common around the time Nickell had been murdered.

Once Britton's profile was complete, police cross-referenced the details against those of the men they had already interviewed and got a strong match – an unemployed man from Roehampton named Colin Stagg.

Stagg was arrested on 18 September 1992 on suspicion of the murder of Rachel Nickell, but there was insufficient evidence to charge him. The police team were, however, completely convinced that they had the right man and launched an ill-advised covert operation – later described as a 'honey trap' – in order to make him implicate himself in the crime.

Using the pseudonym Lizzie James, an undercover police-woman from the Metropolitan Police's Special Operations Group – SO10 – contacted Stagg, posing as a friend of a woman with whom he used to be in contact via a lonely hearts' column. She attempted to obtain information from him by feigning a romantic interest, meeting him, speaking to him on the telephone and exchanging letters containing vivid sexual fantasies.

During a meeting in Hyde Park, the pair spoke about the Nickell murder, but Stagg later claimed that he had played along with the topic only because he wanted to pursue the romance.

'Lizzie' eventually won Stagg's confidence and drew out his violent fantasies, but Stagg did not admit to the murder. During one recorded conversation, Lizzie told Stagg that she enjoyed hurting people, to which Stagg replied, 'Please explain, as I live a quiet life. If I have disappointed you, please don't dump me. Nothing like this has happened to me before.'

When Lizzie went on to say, 'If only you had done the Wimbledon Common murder, if only you had killed her, it would be all right,' Stagg said, 'I'm terribly sorry, but I haven't.'

According to the psychologists monitoring the operation, the more Stagg denied any involvement in the crime, the more clever, cunning and intelligent he was proving himself to be.

Every police investigation is unique and depends on the circumstances of the crime, the evidence available and the resources committed to it. The thing that most investigations have in common is the standard approach the police take. Each investigation is based around the systematic elimination of individuals until only one suspect is left – the offender themselves. Using the acronym TIE – trace, interview, eliminate – every line of inquiry is followed with a view to seeing if the suspect concerned can be eliminated from further inquiries. And so it was that, despite failing to attend appointments and failing to provide a DNA sample, Robert Napper was, on 24 October, officially eliminated as a suspect in the Green Chain rapes.

The reason was simple – various parameters had been entered into the computer database and only those who matched the requirements remained active suspects. For example, in a case where the suspect in an assault has been reported by eye-witnesses to be white, any non-white men whose names are put forward can immediately be eliminated.

In the case of the Green Chain rapes, all but one of the eyewitnesses had reported that their attacker was no more than six foot tall. With eyewitness evidence being notoriously unreliable, police chose to ignore the statement of the one victim who estimated the height of her attacker to be six foot three and removed all suspects in the database who were over six foot.

Napper was taken off the radar and, as a result, no further requests for him to attend a police station or provide a DNA sample were made. His name never came up as part of the Rachel Nickell investigation but, even if it had, the police were already convinced that Stagg was their man.

Detectives may talk about following 'hunches' or putting in hours of legwork, and forensic scientists about the extra care and attention they've paid to particular exhibits in order to find the single scrap of evidence that solved the case, but the truth is, pure luck often has a major part to play in the solving of criminal cases.

Just a few days after being eliminated from the rape inquiry, Robert Napper was arrested as part of a completely separate inquiry into claims that he had been making copies of Metropolitan Police notepaper. His home – a rented room in Plumstead – was searched, and a firearm, ammunition and several knives were found.

Inside a padlocked, battered toolbox covered in peeling red paint police discovered a torch, a restraining cord and medical notes on how to torture people. There was also an illustration of a neck showing how the various muscles in it work and interact. Another showed the anatomy of the human torso.

One handwritten note had written on it 'Mengele's way' – an apparent reference to the Nazi doctor who practised surgical and psychological experiments on both living and dead victims. Police also found a dictionary in which Napper had marked the following words with an asterisk: carcass, Hades, holocaust, immolate, necropolis, regicide, Valkyrie.

The toolbox also contained an *A–Z* mapbook in which Napper had marked the site of each of his attacks or pre-attack surveillance operations with thick black dots. He made brief notes next to some addresses. One was 'sodden filthy bitch'. Another seemed to suggest ways of restraining his victims and ensuring that his DNA was not left at the scene: 'cling film on the legs'. Officers noted that some of the locations in the book were close to those where the Green Chain rapes had occurred. Napper claimed the marks were simply doodles or points on his regular jogging route.

In yet another bizarre turn of events in a case full of bizarre turns, one particular mark on the *A–Z* mapbook, similar to those Napper had admitted making, was on Wimbledon Common, and this immediately made detectives think of the Rachel Nickell murder. However, when the mark was checked it emerged that it had not been made by Napper at all, but was instead a printing error by the publisher.

With little else to go on and no legal power to compel Napper to provide a DNA sample (at the time this could only be done in connection with rape or murder and they had no evidence that

Napper was involved in either), the police team chose to focus their attention on the most serious offence uncovered during the raid – possession of a firearm. Napper was subsequently convicted and sentenced to eight weeks' imprisonment. In court, references were made to his disturbed mental state and a psychiatric report was produced saying he was 'without doubt an immediate threat to himself and the public'.

No further investigation into the other items found at the bedsit took place. Had samples of paint been taken from the red toolbox and analysed, they would have been found to match the tiny flecks that had been found in the hair of Alex Nickell.

In the years since Alec Jeffreys had first developed DNA fingerprinting, there had been a steady stream of progress in refining and improving the technique. This had made it more sensitive and easier to use, but there were still a number of shortcomings. Creating profiles was still a slow, labour-intensive process, so investigations and prosecutions were constantly being delayed, especially if there was a backlog of cases to be dealt with.

All individual police forces around the country kept the results of DNA tests they had carried out, but some of the profiles were based on the original multi-locus probing system and others on the newer, single-locus probing technique. Not only were the two completely incompatible – the results of one set of tests could not be compared to the results of another – but what cross-checking did take place had to be done by hand, as there was no way to input the results into a computer.

With concerns about the validity of the statistics used to support DNA evidence still aired in the courts every now and then, the world of forensic science was in a Catch 22 situation. A bigger database would ease statistical concerns and help validate the science but, without validation, permission to set up a national database might not be granted in the first place.

Before any such database could even be conceived, however, there would need to be a new form of DNA profile that lent itself to computerization. This soon arrived, courtesy of Dr Peter Gill, one of the two Home Office scientists who had originally worked along with Alec Jeffreys to develop forensic applications of the original technique and who had since become the head of research of the FSS.

Gill wanted to find a way to get the polymerase chain reaction (PCR) to copy longer strings of DNA – not just the HLA-DQ alpha gene – so that the technique could be fully combined with single-locus probing. The problem was that PCR worked only with short strings of genetic material, not the lengthy strands that were part of RFLP analysis.

The breakthrough came when Gill identified shorter repeating strings of DNA that could be used as markers. Variable Number Tandem Repeats were between 20 and 100 base pairs long, but the new fragments, known as short tandem repeats (STRs) were between two and six base pairs long – the perfect length for PCR amplification. This length reduction instantly resolved one of the problems encountered in single-locus profiling – the difficulty in analysing degraded DNA.

PCR amplification of STR fragments was introduced in FSS casework in 1994 and was used to analyse four different STR fragments simultaneously. Following a series of improvements and refinements, each adding more STR fragments to the analysis, the FSS developed the DNA profiling standard, known as SGM+, which was able to produce profiles with a discriminating power of one in 1,000 million – a vast improvement on the original SLP systems.

Other advances in the field of electrophoresis combined with automation meant that the size of individual bands in a profile could now be measured more precisely. Previously, analyses by two labs in different parts of the country, or even in one lab on different days, would be likely to produce slightly different results. Automation eliminated these discrepancies and made it possible for each profile to be identified purely by a sequence of numbers. And, unlike before, strings of numbers could easily be stored on a database.

These advances, readily adopted by European Network of Forensic Science Institutes directors, paved the way for the development of international guidelines and validation of commercially produced reagents and instruments to allow accurate, reliable and reproducible interpretation of DNA profiling world-wide.

STR profiling rapidly and totally replaced multi-locus and single-locus probing and SGM+ became the technique used by forensic scientists throughout Europe in criminal investigations.

Being Jewish, I was always happy to work and let my colleagues enjoy time with their families during the festive season. Over Christmas 1992, I found myself called out to what appeared to be a murder-suicide at a caravan park in Swanscombe, Kent.

It was a bright but cold Boxing Day when I turned up, forensic case in hand, to a small residential caravan in north Kent, to find two of my female scenes of crime officers looking a bit peevish. They told me that there were two bodies in the caravan, blood everywhere, and they wanted my opinion as a forensic scientist but also for me to take charge as senior crime scene manager.

Having checked the cordon and what the SOCOs had already done, I donned my one-piece paper oversuit, latex gloves and plastic overshoes and entered the scene. The caravan consisted of a lounge/kitchen and two very small bedrooms. In the first bedroom I found the body of a portly woman of about 18 stone

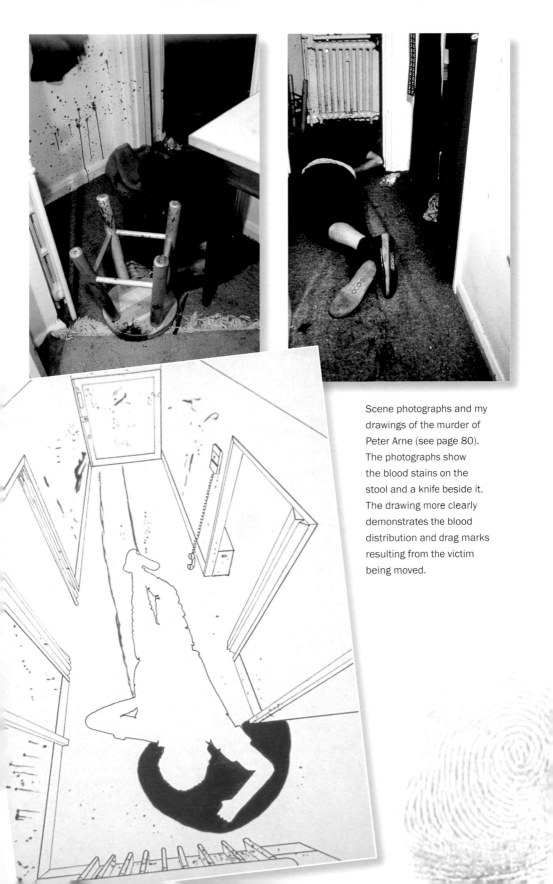

Scene photographs and my drawings of the murder of Peter Arne (see page 80). The photographs show the blood stains on the stool and a knife beside it. The drawing more clearly demonstrates the blood distribution and drag marks resulting from the victim being moved.

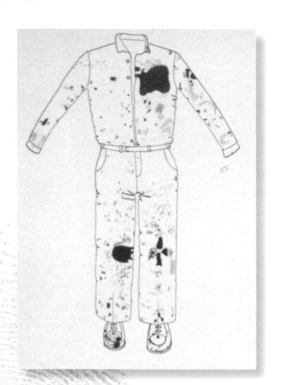
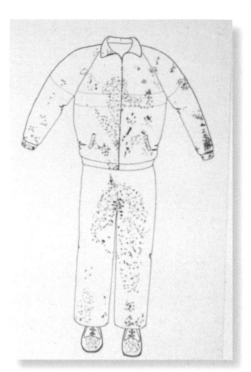

ABOVE: My drawing of the two sets of clothes worn by the killer in the Arne murder. Having changed clothes, the killer realized that Arne was still alive and resumed his attack, resulting in a second set of bloodstained clothing.

BELOW: My plan drawing of the Arne murder scene, showing various sites of blood splashing.

Using a shop-window dummy to disprove the London tower block assailant's claim of self-defence. The blood patterns on his suit indicate he was standing behind the sofa and attacked while his victim was lying down (see page 84).

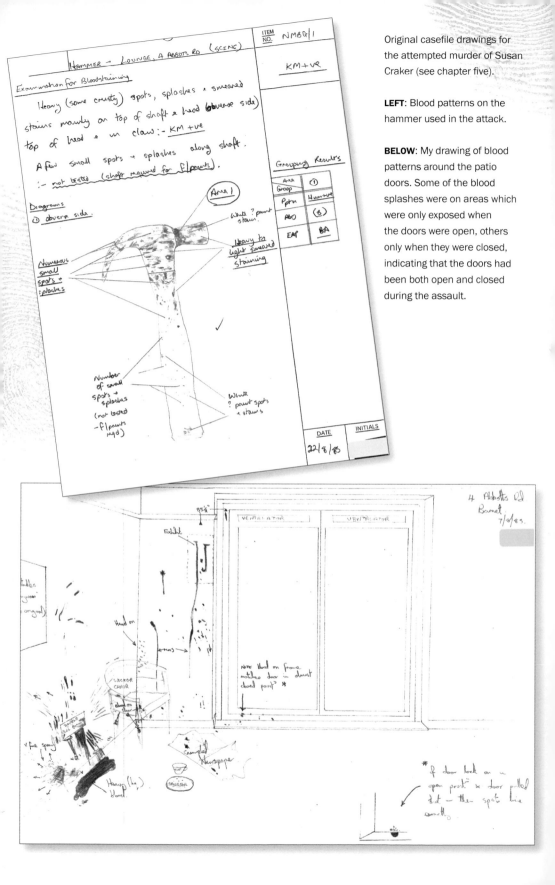

Original casefile drawings for the attempted murder of Susan Craker (see chapter five).

LEFT: Blood patterns on the hammer used in the attack.

BELOW: My drawing of blood patterns around the patio doors. Some of the blood splashes were on areas which were only exposed when the doors were open, others only when they were closed, indicating that the doors had been both open and closed during the assault.

ABOVE: The dye vats in Portugal that proved useful in determining the rarity of the dye in textile fibres in the London rapist case (see page 100).

RIGHT: The tip of a broken knife is an exact fit to the rest of the blade (similar to that in the case of Andy Otto; see page 133).

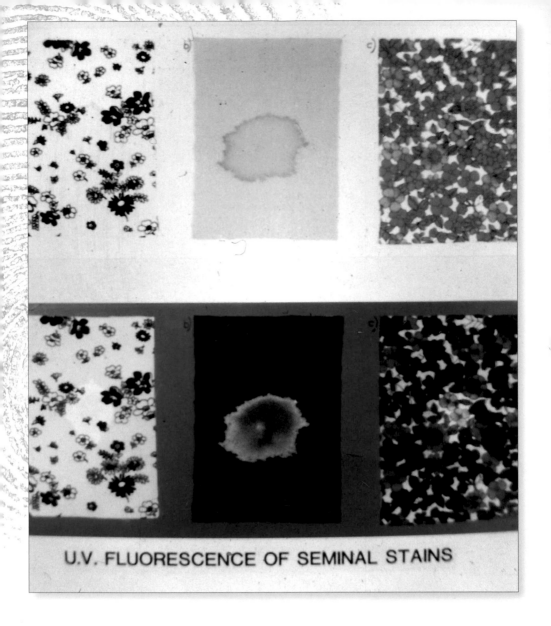

U.V. FLUORESCENCE OF SEMINAL STAINS

ABOVE: Semen stains on three different materials, showing its fluorescence under ultraviolet light.

RIGHT: Testing underwear for semen using acid phosphatase reaction – the resulting purple reaction indicating the presence and position of the semen. This technique was used in the Babygro case (see page 157).

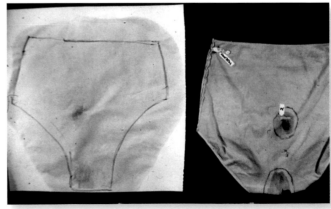

String reconstruction in the Kent murder (see page 223). Trajectories of blood splashes are tracked back to their points of origin and marked with string. The point at which the strings cross is the initial site of impact.

LEFT: Ethiopian forensic scientists carry out blood pattern analysis training under my instruction.

MIDDLE: On board HMS Invincible to audit the ship's rape examination suite.

BELOW: A particle of firearm discharge residue as seen under a scanning electron microscope – which also provides an elemental analysis of the particle identifying its unique origin.

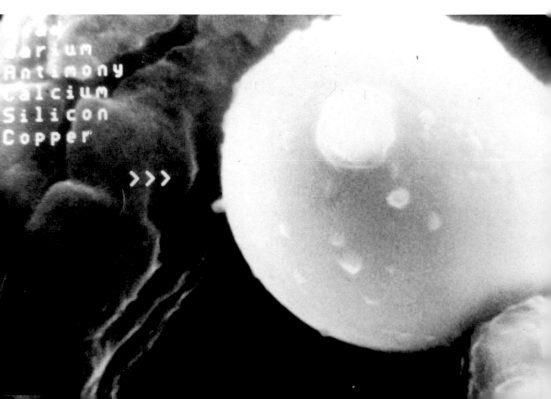

lying on the bed wearing a nightdress, which had been pulled up to her hips. The bedclothes were heavily bloodstained and I could see a stab wound in her chest.

The second bedroom contained the body of a young man, jammed in a kneeling position beside the single bed with his forehead resting on a bedside table. Again the bedding was heavily bloodstained. I was intrigued by the amount of blood concentrated near the centre of the bedding. The man had stabbed himself in the throat by holding a knife to it and bashing his face down on to the bedside table. The knife was still in place, but I could, as yet, see no other injuries.

There was no evidence of blood splashing to indicate a sequence of events, the medical examiner (police surgeon) had already attended to declare that the two individuals were dead and the undertakers had been called, so there was nothing much for us to do but move the bodies out for the undertakers. That's when things started to get interesting.

Rigor mortis had set in, and the woman was like a heavy plank of wood. Her size and stiffness prevented the two slightly built SOCOs getting her to the door, let alone around the corner into the lounge. I took the place of one of them – there wasn't room for three of us – but to no avail. We would have to dismantle the walls. Eventually, and with the help of the senior SOCO, who had now returned, we managed to extract the body. Then it was time for the young man.

The cause of death was clear, but the bloodstaining on the bedding needed an explanation. I had a vague suspicion that the positioning of the staining at the centre of the sheet might have an unpleasant source. I was right. On lifting the man's (much lighter) body I found that he had attempted to emasculate himself – presumably with the same knife.

It was clearly a case of murder-suicide and the police investigator would not be looking for any other suspect, but professional curiosity compelled me to try to reconstruct the sequence of events. The position of the mother's nightdress – for they were mother and son – and the attempted emasculation led me to the proposition that the son may have killed the mother (for whatever reason) and had intercourse with her body, then, perhaps in a fit of remorse, retired to his bed, where he tried to cut his penis off. The remorse, anguish and, no doubt, pain then drove him to take his own life in such a bizarre fashion.

To test this proposition, I had the SOCOs send the autopsy vaginal swabs from the mother to the laboratory for analysis for semen. Uniquely for me, I was right for a second time in one day.

As I was one of the first civilians in the country to be employed in what would normally be a position occupied by a senior police officer, my bosses were never quite sure what I was actually entitled to do. I took full advantage of this by deciding to apply to attend some of the senior command courses at Bramshill – the national police staff training college in Hampshire. I put my name down for a two-week residential Serious and Series Crime course designed to prepare senior officers for dealing with the most serious and complex criminal investigations.

No one objected, so one sunny morning in spring I drove up the mile-long private road in the Hampshire countryside towards the distant red-brick pile with tall, ornate, chimneys. This was the Jacobean mansion that had been bequeathed to the nation's police forces for use as the national training centre. I parked up front on the large gravel drive and went through the imposing front entrance to register. I was promptly redirected 'around the back' – I wasn't anything like VIP enough to park out front.

We had presentations, carried out by the senior investigating officers themselves, of all the more recent high-profile serious crime investigations, including the disappearance of Susie Lamplugh and the brutal murder of Jamie Bulger, and input from a whole host of police units and external organizations that provided resources to the senior investigating officer (SIO), including a presentation from an attractive woman from the FSS on DNA.

The only reason I mention the woman's looks was that, in the discussion the class had after each of the presenters had left, it was clear that none of the police students had understood a word of her presentation. The discussion all turned around whether or not she had been wearing stockings.

'But it was a good presentation,' I offered. 'And DNA is not really that difficult to understand.'

They then challenged me to make them understand it, at the regular evening lecture series where each class member had to present a subject to the rest. As I was familiar with explaining complex science as simply as possible to Old Bailey juries, I took up their challenge and presented 'DNA that Even an SIO Could Understand' the next evening. It went down sufficiently well that the course tutor invited me to replace the FSS in the regular slot delivering presentations on DNA on several Bramshill courses. I accepted and became a regular Bramshill lecturer from that point on.

The highlight of the course was a 'Hydra' exercise, consisting of a two-day series crime simulation, with the whole class allocated various changing roles and each of us taking a turn as the senior investigating officer.

Hydra courses take their name from the multi-headed water serpent of Greek legend that responds to having a head cut off by growing two more in its place. The courses aim to be as realistic as possible and make use of a large support staff that feeds information to the participants in real time. And, just as in fighting a Hydra, each time you reach a point where you feel you have successfully dealt with a problem, two more crop up to replace it.

Two parallel incident rooms were run to make the course into a contest and, although I was officially attached to one of the teams, I ended up acting as forensic science adviser to both. There was no real way to win outright: the teams were judged on how well they coped.

With little experience of this kind of investigation, I resorted to throwing more and more resources at the scenario, with little thought to real-world budgetary constraints. The team I was on lost, but I was given a special certificate congratulating me for being part of the world's most expensive major crime investigation.

I was eager to learn more and, thanks to my new position as a regular contributor to courses at Bramshill, I was able to obtain a place on an even more advanced course that was notoriously difficult to get on to: Management of Disasters and Civil Emergencies.

Once again I made my way up the long drive to the stately home, this time knowing to park around the back, and registered for the two-week course. Again the course consisted of presentations by senior investigating officers and team members who had worked at a number of major incidents, including the Lockerbie bombing. We even visited the reconstructed Pan Am 747, Flight 103.

We were given a deep insight into the myriad problems associated with such disasters, and the SIOs were all brutally honest about the mistakes they had made in the course of their investigations, in the hope that we would learn lessons from them. For example, the ice-rink in Lockerbie had been designated for use as a temporary mortuary and dozens of bodies laid out there. This turned out to be a big mistake, as when the time came to move the victims they had all stuck fast to the ice. No one had thought, until it was too late, that the bodies should have been placed on pallets, not directly on to the ice. A second temporary mortuary site at a local school turned out to be equally problematic. The bodies were being kept in an upstairs hall, but the stretcher-bearers ended up suffering back injuries caused by carrying the victims up and down the stairs.

The finale of the course was another two-day Hydra exercise. We turned up on day one to be presented with a 'live' BBC news broadcast from a large city in the Midlands giving the breaking story of a disaster happening in the city centre. It was very professionally produced, realistic and scary.

We were then taken to a large room containing lots of communication equipment and the trappings of a full major incident suite. In the room next door was a strange sight: a vast model of the city centre in miniature filled the floor, and we were given a few minutes to familiarize ourselves with the layout. As we stood there, the phone in the incident room began to ring. As I was designated one of the two communications operators, I went next door to answer it. The balloon had gone up and the exercise had started in earnest.

I won't disclose the full details of the scenario, just in case Bramshill are still using it, but, suffice to say, everything was thrown in: jumbo-jet crashes, motorway collapses, petrol spillages, people trapped in flooding cars in the canal – and all of this was complicated by a bizarre turn of events which made media handling nearly impossible. After this first day my co-communications operator had had enough and quit the course. 'It's doing my head in' were his parting words.

I regret to say that quite a lot of people died as a direct result of my incompetence. When it was all over, I was exhausted, and the only consolation I could find was that at least it was just an exercise. I came away with an enormous amount of respect for the individuals who deal with such incidents for real.

15

THE SKY WAS JUST GETTING LIGHT ON A CRISP WINTER MORNING when I found myself looking down into the gaping, dank maw of an open grave. The scene was an old but still-used cemetery in Kent; the players, apart from myself, were Professor Taffy Cameron, my senior scenes of crime officer and two of his staff, as well as a senior investigating officer and his detective inspector and sergeant.

Some three years earlier, the young child whose grave we were now attending had died, according to the coroner at the time under no suspicious circumstances, and been buried. More recently, however, her young sister had died – and this time foul play was suspected: poisoning by aspirin overdose, to be more specific. Suspicion had fallen on the child's mother and questions had been raised about the circumstances surrounding the first death. Hence the exhumation.

The graveyard workers had carried out the majority of the excavation of the heavy clay soil in the grave, leaving about six inches of dirt on top of the coffin lid undisturbed. The final stages were of forensic importance, as, in a poisoning case, any residual poison remaining in the body on burial can leech out into the surrounding soil once the body and coffin degrade and decompose. The role of scenes of crime is to recover the coffin and body in as intact a state as possible and to take soil samples from beneath and to each side of the coffin.

'OK,' I said, when all parties were assembled and had been briefed as to what was expected of them. 'Get the remaining soil carefully off the coffin and remove it from the grave.'

I was addressing the senior scenes of crime officer, who relayed the instruction to the two SOCOs. They looked at each other, then back at him, and then at me.

'Oh, for heaven's sake!' I said. 'Hasn't anyone thought to bring a spade to an exhumation?'

They hadn't. Not a spade, a trowel or a spoon. Fortunately, a hardware store was close by and was opening in half an hour. I fished out my wallet and handed the SOCO enough money to buy a spade and trowel and some bacon-and-egg sandwiches to eat while we waited.

This was not an auspicious display of professionalism from my scenes of crime team, but the investigating officer and pathologist were long-suffering, so we all waited. Luckily, the graveside had been screened from prying eyes, so my embarrassment went unrecorded in the local press.

Eventually equipped, one of the SOCOs lowered himself into the open grave, one foot either side of where he expected the coffin to be, and started carefully removing the earth from the lid. In ten minutes the lid was cleared. It appeared relatively intact even after four years. He then started to excavate along the sides. The coffin was virtually flat! The sides had collapsed and pressed the coffin and body into a sandwich about six inches deep.

There was no way the coffin could be raised in one piece without releasing the bottom from the clinging clay below.

Fortunately, this time the SOCO redeemed himself by having in the van a length of piano wire – perfect for such an occasion. Sliding the wire under the bottom of the coffin at the head end, he 'sawed' the wire back and forth between the coffin and the clay beneath it. At last the coffin was released and a tarpaulin slid beneath on which to raise it.

Both SOCOs emerged from the grave and everyone lent a hand to raise the tarpaulin and with it the coffin. As it was a child's, it was not very heavy and it came up quickly and cleanly. It was wrapped in polythene and carried carefully to the waiting undertakers' van, which had now arrived as ordered.

One SOCO jumped back into the open grave and took samples of the soil from several points immediately below where the coffin had lain. We cleaned up, stowed our new digging equipment in the SOCO van and followed the undertakers to Greenwich mortuary for the opening of the coffin and autopsy. We were in for a big surprise.

At the mortuary we all donned one-piece paper crime scene suits, apart from Taffy, who wore the usual green plastic butcher's apron pathologists seem to favour at autopsies. The coffin was cleaned and the lid prised off by the SOCOs, with John, the mortuary manager, assisting.

Taffy had warned us that, in his experience, the body was likely to be severely decomposed and unsuitable for meaningful examination. I was particularly interested, as I had never attended an autopsy on a body that had been buried. Other than being pressed down, the body was in remarkably good condition after four years. The internal organs had been put in a bag at the original autopsy and returned to the body cavity still in their white plastic bag. It was the condition of these organs that would allow – or not – analysis for toxins and poisons to be carried out at the forensic laboratory. We didn't hold out much hope.

Taffy removed the bag from the body cavity, held it over the ceramic dissecting tray that John had provided for him and cut the bag to allow the contents to slide out. And here came the surprise.

I think we were all expecting an unpleasant, unidentifiable liquid to pour out. Instead, the organs were all in near-perfect condition.

'As fresh as the day the child died!' according to Taffy, who exclaimed that he had never seen anything like it in 50 years of performing autopsies.

Nobody could explain the bizarre phenomenon, but I, for one, was grateful that the toxicology examinations were likely to be successful. The samples were taken by the SOCO, sent off to the laboratory and, when the results were received, the mother was charged with two counts of murder.

A few weeks later, on 19 February 1993, two young boys found a biscuit tin containing a Mauser handgun buried on Winn's Common, a large expanse of flat grassland to the east of Plumstead Common in the Royal Borough of Greenwich. A further search of the surrounding area uncovered a second box, this one containing a 'Big Swede' folding-lock knife.

Both the knife and a set of fingerprints on the biscuit tin would later be identified as belonging to none other than Robert Napper, but no further action was taken against him, as he was already in custody for the firearms offence for which he had been arrested a few months earlier.

After just two months in prison, Napper was back at liberty. In July that same year, a middle-aged couple called the police to report a man acting suspiciously in an alleyway close to Winn's Common and Green Chain Walks. The pair explained that they had seen the man spying on their neighbour, a young blonde woman who often walked in her flat semi-naked. The husband followed the man and pointed him out to the police when they arrived. It was Napper. The officers' notes read: 'Subject strange, abnormal, should be considered as a possible rapist, indecency type suspect.' A frisk search showed, however, that he had no prohibited items on his person, so, as he had not committed any crime, he was escorted home and no further action taken.

The extraordinary luck that had accompanied Robert Napper from the start of his criminal career seemed to be holding.

That summer, the Royal Commission on Criminal Justice issued a report with 348 recommendations for the way in which the law enforcement of the future should operate in the UK. Recognizing that DNA profiling had the potential to achieve the same importance in the criminal justice system as fingerprint evidence, the members unanimously supported its expansion.

The commission recommended the setting-up of a national DNA database which would hold details of all those convicted of crimes ranging from burglary to murder. Despite the on-going controversy over its basis in statistics and concerns about keeping on record samples of those who had been found to be innocent, the members of the commission were impressed by the value of DNA as scientific evidence: it was free of the sort of arguments that can surround confessions, statements and witness evidence. The commission also supported greater use of other technology, including closed-circuit television in town centres and more police surveillance equipment to provide court evidence.

In addition to broadening the use of DNA analysis, the commission also called for a loosening of the definition of 'intimate body samples' so that saliva would no longer be regarded as an intimate sample that could be taken only with the consent of the suspect. Reclassifying it as non-intimate would allow police to take samples with the permission of a senior officer. They would have the power to take non-intimate samples, such as hair, for serious offences, even when the samples were not germane to the case, but would not be allowed to use excessive force.

The definition of serious offences would be widened to include not only rape, murder, grievous bodily harm and robbery but also assault and burglary, on the basis that burglars sometimes also commit sex offences and assaults.

The Royal Commission also recommended that when an intimate or non-intimate sample was refused, a court should be able to draw an inference and could take the refusal of the defendant as corroboration of other evidence. Until then, courts had only been able to comment on a refusal to give intimate samples in cases such as rape.

The introduction of the database would require changes in the law, first to allow records to be kept, and second to adjust the caution given to suspects to allow for the inference courts could take into account in the event of their silence or refusal to provide samples.

The Royal Commission recommendations were enshrined in law in amendments to the Police and Criminal Evidence Act (PACE), which effectively gave police the same powers to take DNA samples as they had with fingerprints. Controversial future amendments meant that DNA taken after arrest no longer had to be destroyed on acquittal or discontinuance, and allowed for DNA to be taken after arrest for any recordable offence.

By the time the Royal Commission had issued its report, Robert Napper was becoming increasingly deranged and even more dangerous.

He became convinced that he had been awarded the Nobel Peace Prize and had millions of pounds in the bank. He also believed he had an MA in mathematics and medals from fighting a war in Angola. He told anyone who would listen that his fingers had been blown off by an IRA parcel bomb but that they had miraculously healed after he inhaled special fumes.

Napper struck again on 3 November 1993 at the home of Dundee-born Samantha Bisset and daughter Jazmine-Jemima. Napper spied on Miss Bisset as she made love with her boyfriend before sneaking into her basement flat in Heathfield Terrace, Plumstead, through the open balcony door.

Samantha put up a desperate struggle for life as she was stabbed to death in the hallway. Napper suffocated and sexually assaulted her four-year-old daughter in her cabin bed, surrounded by children's toys. He then set about mutilating Samantha's body in the living room after covering her head and face with clothing. Napper left after arranging her body on a cushion in the position she and her boyfriend had made love. Her partner, Conrad, discovered his girlfriend's body the next morning when he let himself in.

A highly experienced police photographer called to record the scene was so traumatized by what she found that she had not been able to return to work even by the time Napper's trial started.

Fingerprints were taken from the Bisset home, but it took more than six months before these could be fully processed, as it turned out that there were several points of similarity between the prints of Samantha Bisset herself and those of her attacker.

While the investigation was continuing, Napper was still getting into trouble, albeit for far less serious crimes. On 10 January 1994 he was arrested for shoplifting and given a conditional discharge.

It wasn't until 27 May that Napper was finally arrested for the Bisset murders. While he was in custody, a DNA sample was finally obtained and Napper was at last identified as the man behind the Green Chain attacks. On 7 July he was charged with two rapes, two attempted rapes and the Bisset murder. Despite the area in which he had operated and the similarity in the MO – violent attacks on mothers with young children – Napper was not considered a suspect in the Rachel Nickell murders for one simple reason: the case was closed.

After a five-month, £3 million undercover operation, the Crown Prosecution Service had decided that there was sufficient evidence to convict Stagg, and he was charged with Rachel Nickell's murder, despite never having admitted to being responsible for the crime to undercover detective 'Lizzie James' or anyone else. The prosecution also continued despite there being no forensic evidence to link Stagg to the murder.

(Stagg had briefly been considered as a suspect for the Green Chain rapes, and a request was made for him to attend a police station in south London in order to take part in an identity parade, but before the line-up could be put together, Stagg's DNA test results came back, eliminating him from the inquiry.)

Stagg's trial for the murder of Rachel Nickell began in September 1994 but was over almost as soon as it started. As the prosecution presented its case, Justice Ognall ruled that the police had shown 'excessive zeal' and had tried to incriminate a suspect by 'deceptive conduct'. He excluded the entrapment evidence and, with nothing else to go on, the prosecution withdrew its case. Stagg was formally acquitted of the murder on 14 September 1994 but, immediately afterwards, Sir Paul Condon, then Commissioner of the Metropolitan Police, informed reporters that the police were not looking for anyone else in connection with Nickell's murder.

Despite this, the detective superintendent who had led the

Bisset investigation made contact with the head of the Operation Edzell team (as the Rachel Nickell case was now known) to make him aware of the many clear similarities between the two cases. As a result, Robert Napper was interviewed in Broadmoor, where he was serving a life sentence. He denied ever going near Wimbledon Common or being responsible for the murder. With no evidence against him at the time, again no further action was taken.

Police later admitted that, while the senior investigating officer in the Nickell case was aware of the similarities in the two killings, with Stagg awaiting trial, he preferred to await the outcome of the legal proceedings before investigating Napper.

In all, the team investigating the Green Chain rapes identified 86 victims and 106 crimes, but Napper refused to admit any offences for which there was no forensic evidence against him. This meant that, unless DNA evidence could be found to link him to the murder of Rachel Nickell, the case would remain unsolved.

Over at Kent, I had plenty of work of my own to deal with, both in my official capacity as scientific support manager and in my unofficial one as the force's in-house blood pattern analysis (BPA) expert.

Early one afternoon in September 1994, I received a call asking me to attend the scene of a murder that had taken place in a basement flat just off a town centre main street in Kent. Inside the property there were multiple heavy splashes of blood – clearly from several very violent blows – emanating from the area in front of a large desk, which was positioned to look out of the front basement window on to the pavement. The chair was still in position in front of the desk, but the victim's body had already been removed, as was so often the case.

I set about finding the point of origin of the blows and got to work with my magnifying glass, measuring graticule and protractor. There was, however, one change in my equipment. I was now using knicker elastic rather than string to track the splashes and spatters, as I had discovered that this remained taut for far longer and allowed me to mark out the trajectories more accurately – see page 7 of the second picture section for one of my photographs of this process, taken at the scene. To this day, I believe my introduction of the use of knicker elastic to BPA to have been my finest contribution to the field of forensic science.

Measuring the width and length of numerous strategic blood splashes, I calculated the angle of incidence of the flying blood drops that had caused the splashes and marked their trajectories with the elastic. After perhaps an hour and a half I stood up and called the members of the investigating team to report my findings.

'There have been a minimum of four very heavy blows, with the point of impact being approximately three feet above the seat of the chair,' I explained. I had managed to re-position the chair in its exact place at the time of the offence by lining up the bloodstains on the chair leg and the floor.

'Not five feet, then?' asked the senior investigating officer.

'No, three. The victim was sitting in the chair when the blows were struck. Does it make a difference?' Apparently, it did.

The SIO's latest theory was that the assailant was known to the victim. If a stranger had come into the room, the victim would have stood up to challenge him. The fact that the victim remained seated, with his back to the door, in spite of knowing someone was entering, was enough for the SIO to pursue this theory.

And so it turned out to be.

★

The appointment of civilians such as myself in the role of scientific support manager wasn't the only change going on in the world of forensic science. Later in 1994 the forensic science market experienced a further shake-up. Plans were announced to remove three important sources of scientific expertise from the public arena. The National Engineering Laboratory at East Kilbride, the National Physical Laboratory and the Laboratory of the Government Chemist – the last two both located at Teddington – were to be privatized.

With a history that can be traced back to 1842, the Laboratory of the Government Chemist was well-known within forensic science circles. Initially known as the Laboratory of the Board of Excise and based in the City of London, its remit was to regulate the illegal adulteration of tobacco. From there the laboratory took on the role of protecting the government's revenue, its role expanding in 1875 to become the 'referee analyst' under the newly drawn up Sale of Food and Drugs Act. By ensuring that products such as milk had not been watered down, the laboratory not only protected revenue but also looked after the best interests of the consumer.

A century later, the main role of the laboratory was to provide advice on analytical chemistry to the government, and also to businesses. As with the switch to charging by the FSS, the changes took place in stages. For the previous three years (1993 to 1996) the laboratory had been given the target of covering its full costs from income generated, and had managed to achieve this. As a result, the lab, now rebranded simply as LGC, would be turned into a government-owned company (GovCo) before being fully privatized. The FSS now had a serious, determined and professional competitor in the emerging forensic science market.

On 1 December 1996, encouraged by the Home Office, the Forensic Science Service took control of the Metropolitan

Police Forensic Science Laboratory. Metropolitan Police staff were very much against the idea and not too happy at the prospect of working with the FSS. The driving force behind this change, it was claimed, was to create a national service which was demonstrably independent of its police customers.

The biggest problem was that Metropolitan Police staff would have to get used to working in a completely different way overnight. While all the FSS labs had enjoyed a year or two of 'soft' charging in order to accustom themselves to the new regime, the Metropolitan Police Laboratory had no such gentle introduction.

To help smooth the waters, the CEO of the FSS decided to recruit someone who knew the Metropolitan Police Laboratory and had credibility with the staff but also had experience of working with the FSS and of dealing with forensic science as a market service. There was a very limited field of possible candidates, and I was one of them.

I had previously received a phone call from the CEO herself, encouraging me to apply, but initially I wasn't interested. Eventually, however, I did apply and was granted an interview, but I had already decided that I would take the job only if I was able to work on my own terms. I sat down, told the board exactly what I thought was wrong with the FSS, then got up and left.

Within five minutes, just as I was walking across St James's Park to get the Tube to head home, my mobile phone rang and I was offered the job of national account manager with special responsibility for the Metropolitan Police merger. Within a few months, I was behind a desk in the exact same building where I had started my career in forensic science many years earlier.

The hostility between the FSS and the Metropolitan Police was palpable – this was not a friendly merger but an aggressive takeover, according to the Metropolitan Police Service (MPS), who were very unhappy at having their laboratory stolen from them.

To illustrate the pettiness of the lack of cooperation: Michael Howard, the then Home Secretary, was due to tour the newly acquired FSS London laboratory on Monday, the first day after merger, and formally unveil a commemorative plaque. The MPS, who were still the landlords of the property, refused to allow any of the MPS logos, signage or notices on noticeboards to be removed before the handover of the lease the previous Friday evening. However, the FSS insisted that when the Home Secretary toured there should be no trace of the previous MPS management, which meant that all the signs had to be changed at the very last minute.

In the event, all went well, with me and my well-drilled team of five account managers escorting or preceding the VIP party – and tearing down any overlooked MPS notices as we went.

16

JUST AS THE DUST FROM THE TRAUMATIC MERGER BETWEEN THE FSS and the Met Lab was beginning to settle, I was given the opportunity to work on an overseas project, by virtue of my continuing experience delivering lectures on DNA at Bramshill National Police Staff Training College.

The Department for International Development had provided funding to support the development of forensic science in Ethiopia, so, along with my FSS colleague (and former Kent senior SOCO) Bob Green, I headed out there to review the current situation and recommend a strategy for the development of future scientific support.

The project had little money in its budget with which to purchase the latest technology, so I ended up taking a huge amount of redundant blood grouping and other equipment from the FSS which had been inherited from the old Met Lab and superseded by the introduction of DNA analysis.

When the large crate arrived at Addis Ababa airport, customs officials immediately stopped it and indicated that they planned to open it and inspect the contents. This was something of a worry. I had no import papers for the goods and, to the uninitiated, the mixture of lab equipment and chemical reagents might well have been taken for a germ-warfare kit. I had visions of spending a long night in an Ethiopian jail.

Just in the nick of time, the chief of police breezed into the airport with his heavily armed entourage and ordered the customs officials to unhand the crate – it was mine, and hence his, and must be imported unopened. They conceded quietly.

The building where the laboratory was sited was an ex-military installation left over from the Russian/Italian occupation. There, I trained the head of biology, Mulu, and her few staff in the basics of ABO and biochemistry blood grouping. I was present when Ethiopia's first ever electrophoretic blood grouping plate was produced. By strange coincidence, the blood group system used for training was the AK system – the one that had almost made me quit forensic science in the first year of my career, during the Colin Wallace trial.

The Ethiopians had also wanted a DNA database but as the water was too impure to drink (even London water had to be re-distilled for use in DNA analyses) and the electrical supply could not be relied upon for more than five minutes at a time I had to make them settle for something more practical.

I also trained Mulu and her assistant in the fine art of blood pattern analysis. We stood on ladders and benches, dripping goat's blood from a pipette at different heights on to paper held at various angles and measured the length and width of the spots (see page 8 of the second picture section). I made another mousetrap blood distribution device, like the one in the Kent training school, and splashed 'claret' (as blood is often euphemistically termed) around the walls. We battered blood-soaked footballs with iron bars.

Naively, I had expected to train the Ethiopian scientists in techniques that were of value in the UK and so had brought equipment to help them get started in glass analysis for burglary cases. Of course, most homes in the capital had no glass in their windows. I had also brought equipment to help them get started in shoemark analysis. However, I quickly discovered that most of the population went around with bare feet and that those with 'shoes' mostly wore flip-flops made from old tyres.

Once the training was completed, I hoped they would be able to use the laboratory to help prosecutions in sex crimes, which were endemic, though rarely resulted in prosecution. Although both rape and abduction were criminal offences under the law, Articles 558 and 599 of the 1957 Ethiopian Penal Code had provided that, in the event of subsequent marriage to his victim within one year, the perpetrator was exempt from criminal responsibility for his crimes. Payment of a sum to the victim's family, equivalent to a dowry, would also absolve the offender.

This had led to the widespread practice of young girls being abducted by groups of young men and then raped by one member of the group. The elders from the man's village would then apologize to the family of the girl and ask them to agree to marriage. In most cases, the family would give their consent, as a girl who had lost her virginity would be a social outcast and unacceptable in marriage to any other man.

The problem, from a purely selfish forensic science point of view, was that the incident didn't become the offence of rape until the dowry wasn't paid or the marriage wasn't agreed after one year. By that time, it would be a bit late to be taking intimate swabs from the victim.

While I dealt with the laboratory side of things, Bob handled the scenes of crime aspects. This obliged him to travel south from the capital to spend a few days in the country – no easy feat, as Ethiopia had an on-off war going on at the time with

Eritrea. The war was off at that particular point, as the harvest had to be brought in by both sides, otherwise everyone would die of starvation. Bob, however, was advised by his inspector friend and bodyguard to sleep with his feet on the pillow. That way, Bob was reassured, a night-time assailant would only blow his feet off.

When we left the Ethiopian laboratory, the SOCO units were better equipped and trained and had been given a promise of far more aid to come. Unfortunately, Ethiopia went back to war with Eritrea over access to the sea – and the UK pulled out of the project.

As DNA techniques became ever more discriminating and sensitive and the number of samples being analysed and pro-cessed grew ever higher, so the potential for errors increased. Soon after I returned to London from Ethiopia, the FSS found itself on the receiving end of a burst of unwelcome publicity.

Confidential documents showing that there had been at least three cases in which the genetic profiles of suspects had been wrongly matched to DNA taken from crime scenes had been leaked to a Channel 4 documentary crew. It also emerged that, earlier, two DNA samples had been declared to be identical, despite one having been taken from a 14-year-old girl from Merseyside and the other from a 21-year-old man from Lancashire.

The mistakes were all uncovered during routine checking, and the FSS stressed that the system was foolproof, with the chance of a mismatch ever reaching court being 40 million to one, but as the revelations came just at the time the DNA database was getting up and running, they added to the con-cerns of the civil-liberties groups who opposed the setting up of the database.

The database had finally been established in April 1995, and by that time there were 70,000 samples in all, of which 11,000 had been transferred to the system, resulting in 45 confirmed matches.

The mismatch involving the girl and the man was caused by a technician mixing up two test tubes. Instead of analysing two different DNA samples, he examined the same person twice but recorded it as two separate people. Two other errors occurred when bar codes used to identify suspects became corrupted and incorrectly linked suspects with crimes.

Dave Werrett, the Home Office scientist who, assisting Peter Gill, had helped Alec Jeffreys to refine the use of DNA profiling, had been made director of service development at the FSS. He flatly denied that there had been any difficulties with contamination, but it emerged that he had written a memo earlier in the year describing a 'serious breach of professional conduct' at the Birmingham laboratory in which dirty gel from the fourth floor had contaminated DNA samples on the fifth floor.

Human error was one concern, but the ever-increasing sensitivity of DNA profiling itself had the potential to cause problems.

As national account manager of the FSS, one of my key responsibilities was to deal with customer complaints, of which there were many, usually concerned with lost exhibits, invoicing or issues of timeliness. One afternoon in 1997 I had a visit from Hugh Orde, who at the time was a Commander within the Metropolitan Police and would later go on to be Chief Constable of the police service of Northern Ireland and President of ACPO. He explained that we had carried out some analyses for him on a recent murder case and that the results we had provided had led to some major issues.

'We've got a serious problem . . . and I'm pretty sure it's the lab's fault!' he said forcefully.

Analysis of a sample of blood from beneath the fingernails of the murder victim in question had produced a hit on the DNA database, but the troubling part was that the person whose name had been produced by the system had also been murdered, over three weeks before the second victim had died.

It was standard procedure at the time for all blood samples, whether from victims, suspects or merely arrestees to be added to the National DNA Database. If for no other reason, the numbers helped support the validity of the population statistics which the database generated.

Commander Orde's team had immediately checked through their records and compared the two cases. They quickly concluded that there was no connection between the two female victims and no reason to believe one could have had anything to do with the death of the other. The murders had happened miles apart, in different divisions of the MPS, had been investigated by different teams of detectives, and samples had been collected by different SOCOs. So far as the commander could tell, the only place the two cases had come together was at the FSS laboratory, and it therefore seemed to him that, during analysis, one sample had somehow contaminated the other.

It was a reasonable assumption, and devastating for the FSS if it turned out to be the correct one. My job would be to challenge the assumption, identify exactly how the transfer of blood (or data) had happened and devise a system to make sure it never happened again.

Contamination has long been a major area of concern in all forensic examinations, whether at the crime scene or in laboratories, and numerous procedures are in place to minimize the chances of it occurring.

Crime scenes can be contaminated by those investigating and examining the scene. By this time, SOCOs wore disposable paper oversuits, as well as surgical gloves and plastic overshoes at scenes of crime; everyone else, including investigating officers, wore gloves and overshoes. This had not always been the case, and still isn't in some countries – even in Europe.

Contamination may occur within a scene as examiners move about and retrieve samples. This is why photographs and video recordings are taken at the very beginning of a scene examination: to ensure that everything can be seen in its original place. 'Sterile' packaging is used to collect samples and these should be sealed straight away to avoid later contamination.

To avoid possible cross-contamination between scenes – for example, between the murder scene and the suspect's home address – different SOCOs are used where possible, wearing different personal clothing and oversuits, gloves and overshoes. Most SOCOs have personal-issue kits, so different instruments will be used at different scenes to collect samples. Much of the sampling kit is also disposed of after being used at only one scene.

Within the forensic laboratory, laboratory coats and exhibit examination areas are labelled and recorded with a specific reference for each item examined. This ensures that victim and suspect items are not brought together at any point before examinations are complete and extraneous material is removed from the items and preserved. Examination areas and benches are thoroughly cleaned between examinations, and items for examination are often placed on disposable paper sheets. Instruments are scrupulously cleaned or, again, disposable.

My first thought was that perhaps the fingernail clipping from the second victim had been mislabelled and had come from the first victim all along. As soon as I started to look at the samples, however, I could see this wasn't the case. The victim had painted her nails with a distinctive leopard-skin pattern and the cuttings that had been taken bore the exact same pattern, so there was no doubt that these were the correct ones.

I then checked through the laboratory records to see if there was any way the samples could have been switched so that a scientist testing one set of clippings was inadvertently working on the other. This, too, turned out to be a non-starter. The two sets of samples had never been out of the exhibit store at the same time and, in any event, several weeks had passed between the testing of the first and second samples, which had been done by different scientists in different parts of the lab. Naturally, I confirmed the documented evidence by interviewing the scientists involved.

Contamination from examination benches or instruments was also highly unlikely, due to the procedures that were in place and which had been documented on the examination records. After a few days of checking, I called Commander Orde, who came over the river to the lab. I told him that I'd gone through both cases with a fine-toothed comb and could not see any way in which the laboratory could have been responsible. I showed him the documentary evidence to back up my conclusions.

Although I was relieved that the FSS was not the source of the cross-contamination, it was clear that something had happened, and I was determined to get to the bottom of it. I did some further comparison of the cases and realized that both bodies had been taken to the same mortuary for forensic autopsy, although they had arrived there several weeks apart.

The bodies of murder victims can remain in mortuaries for months, and sometimes years, until they are released for burial by the coroner, who has ultimate responsibility for the deceased. This is in order to give both the defence and prosecution teams time to carry out any additional examinations as the investigation continues.

Mortuary procedures for victims of murder differ from those used in cases of death by natural causes. Forensic autopsies determine not only the exact cause of death but also look for evidence of circumstances surrounding the death, such as a struggle or sexual activity. The pathologist will also collect biological specimens, including stomach contents to identify a last meal (and perhaps how long between the meal and death), and blood and organ samples for toxicological testing. Fingernails are scraped and clipped to see if the victim may have scratched their attacker. No such samples are taken during standard clinical autopsies.

When we looked into the mortuary records, we immediately found a possible explanation for what had happened. The body of the first murder victim had been in and out of the freezer on several occasions for second and defence autopsies. Further samples – including nail clippings, taken using a special pair of scissors – had been taken from it the day before the second body was brought in.

Although the scissors had been cleaned in accordance with the standard (pre-DNA) decontamination procedures, they had been used again the following day and may have contained enough minute traces of DNA from the first victim to transfer some to the nails of the second. Invisible to the naked eye, these DNA traces could nonetheless have been picked up by the sensitive tests at the FSS lab, leading to the mysterious link between the two murders.

It was initially only a theory and, in order to put it to the test, I arranged for the nail scissors to be retrieved and examined by the lab. When the analysis came back we found there were DNA profiles from three different people present on them. We followed this up by examining numerous other 'clean' instruments from the mortuary and other mortuaries around the country – and found that almost everything we examined had analysable DNA from two or more different people on it. I later discovered that the FSS research department had already carried out a similar review with the same results, but had failed to tell the customers.

The issue was simply that advances in DNA had made analytical techniques so sensitive that tiny amounts of biological material left behind by the standard cleaning process could be detected. At the time, mortuary instruments were cleaned in a device resembling a large pressure cooker, and this had been more than adequate. With no risk of disease or infection being passed on to dead bodies, this method of cleaning had never been an issue.

It was also a problem unique to the collection of fingernail clippings. Although several of the autopsy knives used to make incisions into bodies were found to have DNA from more than one source on them, such incisions are never sampled for DNA, so cross-contamination would never be an issue. The special nail scissors were only ever used to collect fingernail clippings during forensic autopsies and these were the only samples taken from the body which might then be analysed for DNA. This meant that this kind of cross-contamination simply wasn't a problem with any other part of the autopsy procedure.

Contamination at this level was something that had been postulated in the past, but this was the first time it had been demonstrated in a live case. I put together a laboratory information note that went out to all coroners, mortuaries and forensic pathologists in the country, explaining the situation and advising that, in the future, all nail clippings should be taken with disposable scissors – plastic scissors with metal reinforcement on the blades – and that the scissors should be included in the bag with the nail clippings to confirm that they had been used only once. I went on to give a lecture to the Coroners' Society on the same topic.

Although my laboratory note had no legal or mandatory status, it was nevertheless taken up universally, coming, as it did, from a government agency – the FSS. This was not always to be the case.

At the end of the day, though, it was a really good result. I was happy that we had identified a potential problem early on and stopped it getting out of control, while Commander Orde was pleased to have solid proof that the two cases were not connected after all.

The whole episode put me in mind of a strange occurrence at a murder scene I had attended while still an operational reporting officer at the Met Lab. Attending a murder scene, where, typically, the body had already been removed, I was surprised to find a large sheet of what appeared to be human skin on the kitchen floor near to where the body had been.

The victim had been stabbed and there was no obvious explanation for the anomalous skin. Eventually, after much inquiry, it turned out that the undertakers had been to a fire scene previously and removed a very decomposed body. The same coffin had been used for the bodies, which had first been put in body bags, of the victims of both scenes. A piece of skin had adhered to the underneath of the coffin when it was put down in the fire scene, and had become detached when placed next to the body at our murder scene. Contamination takes strange routes.

The procedure with the disposable scissors is still used to this day and, additionally, other instruments are now cleaned using ultraviolet radiation, which not only sterilizes but also destroys all traces of DNA.

17

THE LONDON OFFICES OF THE FORENSIC SCIENCE SERVICE WERE situated almost directly across the river from the administrative headquarters of the Royal Military Police (RMP), who, one afternoon, paid a visit to the laboratory to find out if we could be of assistance to them.

Members of the Women's Royal Naval Service – the Wrens – had been integrated into the regular Royal Navy in 1993, and the number of women serving aboard vessels was growing fast. With active warships out on operations for months at a time, the RMP believed it was only a matter of time before they found themselves dealing with an allegation of rape or sexual assault.

Keeping the suspect in custody until a full investigation could be started did not present any problems, but as fibre evidence would be key to any case, along with DNA and witness statements, the RMP wanted to know if there were any potential contamination problems. Would, for example, the internal air-conditioning system of a large warship take textile fibres from one part of the vessel and transfer them to another, falsely implicating an innocent party in a crime? In particular, was there any possibility of fibres being transferred between the brig, where suspects would be held, and the medical unit, where victims would be examined?

Furthermore, with all sailors wearing more or less identical uniforms, would it be possible to tell fibres worn by one person apart from those worn by another?

Although I knew a great deal about fibres, I knew precious little about warships. The only way to determine the risk of contamination, I explained, would be to visit one of the ships and conduct a series of tests.

By then, charging for all customers was in full force, but the RMP officers admitted they had no funds with which to pay us. I discussed the matter with my account managers and we all agreed that, from our own point of view, this was knowledge worth acquiring and that we should therefore offer this service as a 'loss leader'.

A date for the work was soon arranged. I'd expected to be visiting a frigate in dry dock somewhere, but was pleasantly surprised to learn that the study would be carried out aboard HMS *Invincible*, the aircraft carrier which, at the time, was the flagship of the Royal Navy's fleet. More than 200 metres long with a crew in excess of 1,000, it was easy to see why the RMP were eager to have such a study carried out.

Along with my colleague John Baker, I travelled down to Portsmouth, where we were taken out to an excellent dinner at the officers' mess. Early the next morning, a Sea King heli-copter arrived to take us out to the ship, which was doing speed trials in the Solent.

We were wearing our survival suits, with the din of the helicopter engine overhead, when HMS *Invincible* came into view. It was a breathtaking sight. Racing along at its top speed of 28 knots, the vessel didn't slow down one bit. Instead, the helicopter pilot had to match his speed to that of the carrier in order to land. It was all hugely impressive. (See page 8 of the second picture section.)

Safely on board, John and I began to conduct our tests. We seeded the air-conditioning system with special fibres that would show up brightly under ultraviolet light, making it easy for us to follow them wherever they went. We then travelled around the ship, placing adhesive tape over various inlets to see if any of the fibres travelled. We knew the system had a series of filters inside, but no one knew how effective they were against fibres or whether there were certain routes particles could take to go from one part of the ship to another without being trapped on the way.

Once those tests were completed, we also visited the rooms where victims would be examined and suspects held, gathering all the details of exactly what procedures would be used. We would later conclude that none of the fibres had been able to travel across the ship and that, so long as certain precautions were taken, the chances of contamination were extremely low.

Once we'd finished, we took a Sea King back to Portsmouth for another night at the officers' mess, before returning to London. An in-depth report was written and sent to the commanding officer of the RMP, confirming our findings. Our work had produced another satisfied customer but had not generated any income for the organization, and, increasingly, the pursuit of profit was the driving force behind every decision the senior management of the FSS were making.

It was believed that the emergence of a more competitive forensic science market in the UK might allow the police 'customer' to drive the use of forensic science through direct charging. It was hoped that opening the market up would push prices down and efficiency and customer service up through open competition between the government-owned FSS and other private suppliers. Naturally, it never happened quite like that, and there were unintended consequences.

Although I was enjoying my position in the FSS, the move to a more commercial 'culture' was having a profound effect on the internal structure of the organization, as well as on the attitudes of both staff and customers.

The police had told me, in my role as national account manager, time and time again, that what they wanted from the FSS was maintenance of services, a quicker turnaround of cases, delivery to set target dates – all as cheaply as possible. Competitors such as LGC were already offering much of this, and in all likelihood were often making a loss on some cases simply to build up a reputation for efficiency and cost-effectiveness which would guarantee them future work. If the FSS didn't do the same, it was going to lose out.

Part of my job was to tell management what the police 'customer' wanted from the organization, but the FSS wanted to push new ideas and technology and saw itself in the future taking over crime scene examination roles and fingerprint services.

The FSS scientists had performance targets imposed on them, based on meeting delivery dates (good) and individual income generation targets (not so good). Consultants hired by the FSS had introduced the US business principle of 'supply chain' management. The huge backlog of cases was re-branded as a 'forward load', which didn't sound so bad, and later as a 'full order book', which sounded positively good, but it was still a backlog to the police, who saw through such management-speak.

Target delivery dates were an important and fundamental issue and one which the whole concept of a competitive market was supposed to address. The problem was this: when a case was brought before the court by the CPS, the forensic science evidence (if any samples had been submitted to the laboratory) was invariably unfinished, and the CPS would ask for an adjournment to await the scientific results. The court would set a date by which the work had to be produced. The CPS would inform the FSS, who would agree to use this as a target date, or explain why more time was needed and re-negotiate.

So far, so good. But the case, once submitted to the FSS, would often sit for weeks or longer in the backlog before being allocated to a reporting officer. The examinations and analyses would take a fixed time and then the case would remain in the RO's out-tray until the statement had been written. Often the target date would be missed. The level of inefficiency – the direct result of those in the organization having spent so many years working in a total monopoly – was becoming increasingly unacceptable.

This situation had been reluctantly tolerated by the courts for many years, with the CPS having to go back and ask for another (expensive) adjournment, and with the suspect often being remanded in custody. The change came when new legislation set custody time limits requiring the release of the suspect after a specific number of days of arrest or charge. The courts and the CPS were no longer going to be so understanding, and the police would be held to account for the punctual delivery of case evidence.

Now target dates were serious currency. Police forces wanted their money back if agreed dates for delivery of test results were missed, and failure to comply meant that the forces were far more likely to take their future business elsewhere. As national account manager, I received complaints from irate investigating officers and, to add to the problem, discovered that some ROs were manipulating the system to hide missed delivery dates.

The situation was particularly alarming in the light of forthcoming legislation, the 1999 Local Government Act, which placed a statutory duty on police forces to ensure that they received the best possible value for money for the goods and services they procured.

The FSS had no official contracts with any of its police customers – instead, everything was based on a series of 'gentlemen's agreements' – but with few other places to go (the FSS had 80 per cent of the market at this point), there had never been any need for concern. However, the Local Government Act meant that whatever semi-cosy relationship between the police and the FSS still existed was about to come to an abrupt end.

In an attempt to meet financial targets and sharpen commercial focus among staff, personal income-generation targets were established. Having experienced the tendency of some reporting officers to get carried away and turn cases into major research projects, I did my best to look out for inappropriate analyses, but a proper clampdown would have been possible only if I had double-checked every single case invoice – an impossible task.

The focus on income did, however, make me suspect that there were times when staff would use advanced, more expensive DNA analysis techniques in preference to standard cheaper ones whether the case required them or not, simply in order to be able to generate a greater level of income. The preference for such techniques would lead to a scandal that would ultimately contribute to bringing about the end of the FSS itself.

I became increasingly disillusioned with my position at the FSS. Senior managers were not interested in the needs of the police customer, other than at a strategic level. There was an unacceptable degree of interference from the marketing department, which was trying to dictate the nature of my work, and a growing culture of 'not my problem' among the staff – engendered by the 'supply chain' business model.

Take the Alice Rye case, for example. The 74-year-old church volunteer was murdered in her home in the Wirral in December 1996. The shocking level of brutality employed by her killer ensured the case made headlines around the country and that police were under huge pressure to catch whoever was responsible, particularly as other pensioners in the community were living in terror as the killer remained at large.

Exhibits from the case were submitted to the lab closest to Greater Manchester Police – Chorley – but because of a heavy workload, some elements ended up in London, including the victim's wedding ring. Having examined the precious object – only Ms Rye's blood was found on it – the reporting officer assigned to the case promptly lost it.

The ring was, understandably, of huge sentimental value to Ms Rye's family, who were deeply upset when it failed to be returned to them along with the other items that had been submitted for forensic examination.

The complaint ultimately arrived at my desk. When I questioned the reporting officer responsible, it was clear that he could not have been less interested. His position was that he had had a quick look for it, couldn't find it in the places noted in the file and therefore didn't know where it was. He felt that the problem was now mine and mine alone.

Frustrated at his lack of cooperation but eager to do everything I could to assist the family, I press-ganged all five of my account managers and some half-dozen members of staff from the customer service department to form a full-scale search party. We turned the whole laboratory over and must have spent twenty man-days looking for this one small item. It eventually turned up: in the desk drawer of the reporting officer.

The reporting officer remained thoroughly unrepentant, saying that it was still for me to deal with, until I told him that I had arranged for him to attend a face-to-face meeting with Alice Rye's family so he could explain the situation. I quickly received a grovelling apology, happy that he hadn't called my bluff.

The case had a successful conclusion, but not for the FSS. The development of the market in forensic science, combined with the statutory demand to obtain best possible value meant that, for the first time, police forces were able to obtain a second opinion about exhibits.

Forensic scientists from LGC re-examined female clothing found at the home of the chief suspect, police informant Kevin Morrison, and were able to create a profile linking the items to Alice Rye. Morrison was jailed for life.

Yet another failure to see eye to eye with the marketing director in the summer of 1999 led to my resignation in favour of a new role as head of international business for the FSS. Having enjoyed taking part in the project in Ethiopia, I relished my new position and was even happier when the Chief Executive made it clear that, if she ever saw me in the UK, it would be taken as a sign that I wasn't doing my job properly!

The United Kingdom was an active participant in projects supporting candidate countries in their preparations to join the European Union, its contribution being managed through the Foreign and Commonwealth Office. The FCO and central government had a high regard for these projects and took a keen interest in promoting the participation of government departments and agencies. In 1999, only the second year of European Union funding, the UK had been involved in only a very few projects in the field of Justice and Home Affairs.

Eager not to disappoint, I spent the next three years

travelling around the world, selling FSS expertise and services to countries that were keen to develop forensic science facilities of their own. My travels provided a remarkable insight into the way the criminal justice system works in other countries.

The FSS was at that time recognized as a world leader in forensic science and had an unrivalled capacity to draw on highly specialized forensic skills and knowledge. Many of the scientists were acknowledged throughout Europe, and beyond, as world authorities in their disciplines. Here, then, was a ready source of international income and funding that would not only help secure the financial future of the organization but also put the FSS strategically in the centre of a high-profile UK government priority programme.

One early project took me to Poland, where the local police wanted assistance in setting up a national DNA database, similar to that in the UK. I was driven around the country along with a DNA colleague in order to attend meetings in Warsaw, Katowice and Opole.

Although not as basic as the facilities in Ethiopia, the laboratories we saw in Poland were still very Soviet and utilitarian in their structure, and staff were concerned about security and snooping. In spite of the modern scientific techniques these countries were eager to introduce, the security systems in place were straight out of the dark ages – but hugely effective.

I had noticed that on each office doorjamb there was a short length of coloured string, and on the door just adjacent to that was a small circular brass ring with a slot through it. On leaving the office, the manager would close the door, bring the string through the slot and apply a small piece of plasticine to the raised brass ring. They would then use their personal signet ring to press a unique seal into the plasticine, which sealed the door. It worked perfectly.

A colleague, Andy Williams, joined the team, and together

we secured an FSS first, with a two-year European Union funded project delivering forensic science development, including a DNA database, to the Czech Republic. Andy, based in Prague, was given an office, which we ascertained was 'bugged'. We had fun then, having rehearsed conversations designed to bring the Czech lab director running to Andy's office to offer advice.

The project went well. I went to Prague every few months for strategic meetings, and lots of FSS staff went to Prague for short-term attachments. After two years the Czechs had a better DNA database than we had in the UK, and we were looking for a new EU project. Fortunately, the European Union funding programme was still running.

Andy Williams was ready for our next long-term project — in Lithuania. The structure was similar to the Czech project but this time the Lithuanians required a whole new laboratory to be built, equipped and its staff trained in the complete gamut of forensic science activities — including the installation and use of a DNA database.

Along with my team I spent the best part of two years cajoling the Lithuanian police, finance ministry and forensic scientists into knocking down and rebuilding a brand-new laboratory complex. With the help of experts from the FSS who came over on short contracts, we advised on and arranged the purchase of the latest forensic science equipment, computers, microscopes and DNA technology. We then organized comprehensive training of all of the Lithuanian scientists at the facility in their new equipment.

I helped out again with blood pattern analysis training, decking out a shower block with brown paper sheets and using the shower cubicles to batter bloodstained objects with a variety of weapons. Splashes were measured and string reconstructions built. A pump was set up to simulate arterial spurting when a tube protruding from the top was cut. I built yet another mousetrap device. The half-dozen or so Lithuanian scientists were joined by their senior SOCOs for the training.

The project was particularly gratifying, as the country ended up with a forensic science capability that was second to none and, in recognition of my contribution, I was awarded the honour of Guardian Angel (First Class) – a medal presented by the Chief of Police, and a first for a non-Lithuanian civilian. Andy got a watch.

While the long-term EU-funded projects were running, we had time to deliver a number of short 'scoping studies' to try to lay the seeds for future business for the FSS.

The nature of my work meant I often found myself travelling under armed escort or in areas that were effectively war zones. In Algeria I wasn't allowed to go anywhere except as part of a five-car armed convoy; I asked to visit a SOCO office and they closed the entire bazaar in the centre of Algiers for two hours and had police snipers on the rooftops the whole time I was around.

In the Gaza Strip, I visited a laboratory which had been built from local stone and in which the chemistry benches were made of marble. I tried to point out that this was the worst possible material, as it would dissolve if acid was spilled on to it, but it turned out not to be an issue. A few weeks after I visited the laboratory, it was obliterated in an Israeli air raid.

Having survived two such trips, I had no wish to exceed my quota of luck. When an invitation arrived to carry out a short scoping study in war-torn Sierra Leone, I sent an assistant. Instead, I travelled to the British Virgin Islands to look at the options for developing a comprehensive forensic science facility there.

The international work didn't always go so smoothly. I arrived in Serbia one year, once again to provide advice on setting up a DNA database. The trip began badly when government officials insisted on taking me to visit their laboratory facility, which had been destroyed by British missiles during the Bosnian war, so that I could witness the destruction first-hand.

When we finally got around to having meetings about the database, the Serbians were asking a number of highly detailed questions about just what information could be gleaned from DNA profiling. In particular, some were keen to know about race-linked genetic markers.

It soon became clear why certain Serbians wanted a DNA database. It didn't have anything to do with fighting crime; they wanted it for the purposes of ethnic cleansing. I made my excuses and returned to the UK as quickly as possible, and withdrew from the consultation at once.

18

BY THE SUMMER OF 1999 THE MARKET FOR FORENSIC SCIENCE was becoming increasingly combative. The FSS had become a trading fund, which meant that the service was allowed to use the money it earned from selling its services to fund its own outgoings. It was the first step in the long road to converting the service into a private company. One consequence was that there was more pressure than ever to compete against the growing number of private companies which were also offering forensic services to the police. Luckily for the FSS, the company had an ace up its sleeve.

That same year it introduced a stunning innovation into the world of forensic science. Known as 'low copy number' (LCN) DNA, it offered remarkable new levels of sensitivity, providing the ability to generate full profiles from just a few cells of biological material, even with degraded samples.

It was a stark demonstration of just how far the technology had come in a relatively short space of time. A decade earlier, you needed a sample of reasonably fresh blood or semen the size of a five-pence piece to get a DNA profile, and it would take two to three weeks. As the years passed, the sample size required reduced and the time needed dropped until it took less than 48 hours to produce a profile.

The development of LCN DNA seemed to be the next logical step, allowing a profile to be produced even where none of the usual body fluids – spermatozoa, saliva or blood – was present. With LCN, a profile could be obtained from an object which a suspect had merely touched.

At the time, the standard DNA profiling technique, SGM+, was limited to the optimum performance of 28 cycles of PCR and worked with as little as a single nanogram of DNA. This equates to 160 human cells – the amount you might find in the smallest spot of blood visible to the human eye. Although 28 cycles worked effectively with minute traces of body fluids, those cases where there was even less DNA, or just one or two human cells, failed to generate a result. This was where LCN DNA came in.

Research conducted by the FSS found that by increasing the number of amplification cycles from 28 to 34 and optimizing other parts of the process, a full profile could be obtained from DNA samples that, previously, would have been too small. However, there was a problem: increasing the number of cycles boosted the amount of DNA, but it also led to a range of issues that could compromise the final results. Although the technique resulted in more positive profiles, it also greatly increased the possibility that DNA unrelated to the case under investigation – perhaps from an innocent party, or a technician involved in conducting the analysis – might also be amplified and detected.

Another difficulty with LCN DNA was that, because the technique was specifically designed to work in cases where a minute amount of DNA was present, excess DNA in the original sample could have an adverse effect on the result. If sufficient DNA was present to carry out a standard SGM+ analysis, using LCN DNA analysis could be prejudiced, due to contaminants in the sample interfering with the reaction and failing to produce a result. This is known as PCR inhibition.

The usual way to avoid this is to carry out quantification – measurement – of the original sample to see how much DNA is there, and diluting it if necessary. But this takes time, makes the process slower, risks using up some of the precious DNA in the original sample and is more expensive, as a commercial kit has to be used.

Inhibition can easily be avoided by just diluting a sample. A PCR analysis would be duplicated at several increasing dilutions to ensure that any inhibiting agents were diluted out. However, duplicating the analysis means using more reagents and more time, all of which raises cost.

Other forensic providers soon began offering their own versions of LCN, more correctly called low-template DNA (LTDNA). A major difference was that these other providers would routinely quantify samples to see exactly how much DNA was present prior to running the amplifications. The other difference was that, rather than carrying out 34 cycles, they would perform 28 cycles and then submit the product to a 'clean-up' procedure prior to analysis by optimized capillary electrophoresis. This process purified the sample and prevented the effects of inhibition from being seen. It produced profiles which were directly comparable with those of the FSS without requiring additional amplification.

Forensic scientists generally split their limited amount of DNA into two or three samples and run analyses on two of them. The third, if available, is reserved for the defence. The results of analyses weren't completely reproducible, profiles often would not match exactly and scientists generally accepted as true those STR signals that showed up in both runs.

There was also a danger of bias errors. It was easier for forensic scientists to distinguish between 'noise' and what had really come from the DNA if they had a reference sample to work with. Low copy number or not, that reference sample was often the suspect's DNA – and faint peaks in a crime scene sample might seem more convincing when viewed side by side with a strong peak from the suspect. The fingerprint world had learned lessons from similar biases.

Usually, when a new scientific technique is unveiled, it undergoes a process of validation, first by other workers at the company which originally developed it and then, after publication in scientific journals, through review and scrutiny by the rest of the scientific community. However, commercial pressures were building up at the time and the management of the FSS was keen to have an edge on other companies in the field, particularly newcomers to the private sector such as LGC. A technique which promised to work on thousands of cases in which standard DNA techniques did not was a potential goldmine. The problem with LCN DNA was that the modifications were relatively minor and it would be easy for any other company to duplicate the technique to achieve the same results.

As LCN offered a greater level of sensitivity, it consumed more resources and required more in the way of preparation, so the technique cost more – but it could also be sold for a higher price than the standard test. A combination of these pressures meant that, by the time LCN was being offered to police forces and employed in cases that were making their way through the courts, the technique had not been validated by anyone other than the FSS itself.

It was a repeat of the situation experienced in America a few years earlier, when Cellmark and Lifecodes were racing one another to bring their products to market by missing out similar levels of validation – a situation that almost led to DNA evidence being excluded from the American judicial system. It was surely only a matter of time before the adoption of a similar approach backfired for the FSS.

The first use of the new technique was, however, a complete success. In August 1977 Mary Gregson was raped and strangled then dumped on the towpath of a canal in Leeds. More than 11,000 people were interviewed by police, including every one of the 800 men working on the construction of new Inland Revenue offices close by.

One man strongly considered to be a suspect was Mary's own husband, Bill. Considered a little strange by some locals, his blood group matched that found in forensic samples on Mary's body.

He made television appearances, saying that he knew that his wife's killer was local and would soon be caught. But even that did not remove the deep suspicions of both police and his neighbours. The finger of suspicion also pointed at all four sons of one Shipley woman. The police search of her house resulted in her being shunned by other women in the town, and her life became a 'living hell'. 'All my sons were questioned, my house was turned upside down and we were followed everywhere we went for at least 10 days,' she said later. The murder tore the town apart.

With no witnesses and nothing to go on, the only evidence West Yorkshire police had was the crucial forensic samples taken from Mary's underwear. Even the 200 various fibres lifted from her body did not help. They could as easily have come from the mill where she worked or the river in which she was dumped as from her killer's clothes.

By late 1988 police worldwide were employing DNA analysis to solve crimes. The West Yorkshire murder squad re-opened the Gregson files and waited for three weeks for forensic scientists to complete their analyses.

It ended in more frustration for the officers. The samples used had by then broken down so much they could not be matched with those originally taken from the local male population. The murder team refused to give up, though, and in 1994 it pinned its hopes on improved DNA profiling techniques – only to have them dashed again. This time, tiny ingrained particles of chemicals on Mary's underwear – most probably from her job at the mill – prevented accurate testing.

Then, twenty-three years after the murder, the investigating team sent the remains of the semen from the scene to the Forensic Science Service and LCN DNA was used to obtain a profile. The police then obtained voluntary samples from potential suspects, one of whom was labourer Ian Lowther, by then a grandfather. His profile matched, he admitted the crime, and was sentenced to life.

Newspaper reports suggested that Lowther was one of 600 people in the UK at the time who had got away with murder but would be brought to justice thanks to the advances that had been made in DNA – chief among them the introduction of LCN.

As the techniques employed by forensic science became ever more sophisticated and sensitive, so consideration of the relevance of the results became more critical. A murder suspect who is found to be wearing heavily bloodstained clothes – in particular, clothes on which the patterns of staining are consistent with assault on the victim – is strongly linked to the crime. If, however, a single tiny stain of blood is found on the suspect, it is not possible to say with any certainty whether it came to be there as the result of direct involvement in the crime or as the result of secondary transfer.

One of the most startling demonstrations of this emerged

during the investigation into the murder of television presenter Jill Dando, who was shot outside her west London home in April 1999. She was one of the most popular and best-known figures on television in the UK – she had won BBC Personality of the Year in 1997 and had co-presented *Crimewatch* since 1995 – and the pressure on the police to find her killer was intense.

After more than a year, during which seemingly little progress was made in the investigation, an arrest was made. On paper, Barry George seemed a credible suspect. He had a string of convictions going back to 1980, when he had been charged and convicted of posing as a police officer in order to gain entry into the homes of women he had stalked. In the years that followed he picked up convictions for indecent assault and attempted rape, serving 18 months in prison. In January 1983 George was arrested in the grounds of Kensington Palace, home to Diana, Princess of Wales, and Prince Charles, having been found hiding in nearby bushes wearing a balaclava and carrying a knife and a length of rope.

Known locally as something of an oddball, George quickly rose to the top of the list of potential suspects. He lived in the vicinity of the crime scene, had spent time in the Territorial Army and received firearms training, and matched the build and height of a suspect who had been seen shortly after the murder. Furthermore, George had been questioned in connection with the murder of Rachel Nickell and was considered to be someone with the potential to harm women. So far as the police were concerned, all the signs were in place – the only thing missing was hard evidence linking George to the crime.

Clothing removed during a raid on his home was submitted to the Forensic Science Service and, on Tuesday 2 May, senior scientist Robin Keeley began to examine a dark-blue Cecil Gee coat that George had purchased for his wedding back in 1998 and worn rarely since. Keeley removed the item from its sealed evidence bag, placed it on his lab bench and began dabbing various parts of it with a small piece of adhesive tape in order to lift extraneous particles from it.

Over the course of the next few days Keeley examined the samples under a scanning electron microscope capable of resolving particles a few nanometres in diameter. Most of what had been lifted turned out to be common household dust, but one particle stuck out from the rest. It was minuscule – just over one hundredth of a millimetre in diameter, far too small to be seen by the naked eye – and had been taken from the left inner pocket of the coat. It was a particle of firearms discharge residue (FDR). (See page 8 of the second picture section.)

Keeley checked other tapings from the same pocket and elsewhere, but no other particles were found. A year earlier, Keeley had also examined firearms particles taken from the wound in Jill Dando's head. The particle from George's coat was of a similar composition.

This did not mean the particle came from the same gun – many guns would have produced an identical composition of FDR particle, even if they were loaded with blank ammunition, as could residue from a firework. Despite this, the police treated the finding of the single minute particle as the missing piece of the puzzle and, within days, George had been charged and was awaiting trial. The finding of the residue was the main plank of the prosecution case and, in 2001, George was convicted of Dando's murder by a majority of 10 to 1.

This conviction was later quashed, in part due to the fact that the FSS, disturbed by just how much emphasis had been placed on the residue during the original trial, produced a report that stated that, due to the spectre of indirect contamination, they would have been just as likely to find a single particle of similar residue on any member of the public. 'It would be just as likely that a single particle of discharge residue would have been recovered from his pocket whether or not he was the person who shot Miss Dando nearly a year previously,' the report stated.

Despite its initially compelling nature, the finding of the single particle of firearms-discharge residue had a completely neutral effect on the case. With science rapidly advancing to the point where a microscopic particle of DNA could be recovered from a crime scene, it was clear that similar issues were set to emerge in years to come.

19

BY THE SPRING OF 2001 THE POLICE FILES ON THE MURDER of Rachel Nickell had remained virtually untouched since the failure of the prosecution of Colin Stagg some six years earlier. Although the case remained open, little, if any, work had been carried out, chiefly due to the continuing belief among members of the investigating team that Stagg was guilty.

Two independent reports commissioned in the aftermath of the murder of London teenager Stephen Lawrence had recommended the abolition of the 'double jeopardy' rule – a fundamental principle of English law since the Norman conquest, which prevented defendants from being tried twice on the same charges. The old law meant that, even if the police managed to find new evidence against Stagg, they wouldn't be able to do anything with it. In the fallout from the reports, the statutes were being adjusted so that, so long as there was 'new and compelling evidence', a second trial would be permitted.

Well aware of the many advances in DNA technology the Forensic Science Service had made, chief among them the super-sensitive LCN DNA technique, which was proving its value time and time again, the police felt they had good reason to be optimistic. Tapings taken from the body of Rachel Nickell were submitted to the FSS and underwent the LCN DNA process. Much to the disappointment of all those involved, no trace of DNA – either male or female – was found to be present.

In the meantime, the FSS continued to promote other innovations that were just on the horizon. The same month the Nickell tapings were tested, the *Independent* ran a piece on all the new developments based around an interview with Dave Werrett, who was by then director of operations.

Werrett made public that the FSS was just a few years away from being able to provide police officers with detailed descriptions of suspects whose DNA had been discovered at crime scenes, such as skin shade and size of chin, ears and nose. He also explained how their increasingly sophisticated techniques had enabled them to obtain viable profiles from a chewed pen-top, a watchstrap, a cigarette end, old bits of bone, and maggots infesting a corpse.

Werrett concluded the article by talking about one day using computers to compare footage from street cameras to a database of people's facial characteristics. 'A person couldn't do it, but a computer could do it, if it was given the parameters of nose and ear size, and so on. So the computer could get you down from, say, 20,000 faces to 50 faces, and after that you'd go into human intervention. That's the kind of thing we're thinking about.'

The director of operations had always waxed lyrical about new technology. He had given multi-screen presentations to FSS staff about the 'lab on a chip' project – silicon technology that would enable the DNA analysis process to be reduced to a SIM-card-sized silicon chip in a hand-held device.

Werrett also tried to realize a vision of mobile DNA laboratories built into vans which could travel to crime scenes and perform detailed analysis on the spot. Following on from the plans for a hand-held device, the new project was instantly dubbed 'lab in a chip van', but failed to live up to expectations. The technology simply didn't work on the move and stories began to circulate about blood swabs from crime scenes being passed into one window of a mobile DNA van then out of the opposite window and into the hands of a motorcycle courier who would take the sample to a real laboratory. I had heard the story first-hand from several FSS staff who were in a position to know. I was unable to confirm the truth, but I did see the vans depreciating in the car park instead of being out at crime scenes.

The failure of the mobile labs did nothing, however, to halt Werrett's rise through the organization. When CEO Janet Thompson announced her retirement, he was named as her replacement.

As soon as he took over, he set about rearranging the FSS, shifting the focus to the UK rather than international business. It seemed to be an odd decision – increasing competition meant prices for UK forensic services were falling, which in turn meant the size of the market was shrinking, but no one in the organization seemed to see the warning signs.

The international business department – my department – was subsequently shut down, even though it had been making a small but steady profit. It meant that I, quite literally, had nothing to do and spent weeks twiddling my thumbs until I decided to act on an opportunity which was coming up within the Science Policy Unit at the Home Office. It was looking for someone to assist in the development of a set of quality controls and nationally agreed standards for forensic pathology in England and Wales.

Forensic autopsies – those required in cases of suspicious death – could be carried out only by pathologists who were on the official Home Office register of Forensic Pathologists. There were then 44. In fact, the Coroner could appoint anyone they wanted to perform an autopsy but, in practical terms, they almost invariably used pathologists who were on the Home Office register.

The problem was that the criteria for getting on to the register were unclear, as was the level of experience necessary. Worse still, there were no procedures in place for dealing with pathologists who made terrible mistakes or errors, or who were just plain incompetent. Although they could be disciplined by the General Medical Council in extreme cases, taking them off the Home Office register was almost impossible.

A longer-term problem was the fact that there were not enough new pathologists coming through the NHS training schools to cope with future demand. The NHS had underestimated the number of pathologists needed in the future, courses had been cut and very few were coming through the traditional route. Part of the problem was that all the publicity about rogue pathologists and expert witnesses being harshly cross-examined in court was putting off medical trainees. Why go to the Old Bailey and risk your professional reputation when you could be paid exactly the same money to sit in a laboratory and look through a microscope all day?

I was offered a secondment to the Home Office and given a budget of £16 million to develop a national network. I soon realized that this would require me to set up standard criteria for entry to the register and to introduce a disciplinary procedure and quality-management system. Having worked closely with a large number of forensic pathologists during the course of my career, I seemed well suited to the task.

The timing was perfect. Not only was I in dire need of a new challenge at work, but the disciplining of forensic pathologists was just about to become a headline issue.

★

After six years of marriage, Sally and Stephen Clark, both solicitors living in Wilmslow, Cheshire, welcomed their first child into the world on 26 September 1996. Christopher Clark was a normal, healthy baby, but on the evening of 13 December Sally called an ambulance to the family home, explaining that her son had fallen unconscious soon after being put to bed. Stephen was out at an office Christmas party at the time and, when the ambulance arrived soon afterwards, Sally was so hysterical she was unable to unlock the front door.

When the paramedics finally reached the child, he was displaying symptoms of cyanosis – the bluish discoloration of the skin caused by a lack of oxygen. He was pronounced dead at 10.40 p.m.

An autopsy was carried out by Home Office-registered pathologist Dr Alan Williams, and the death was subsequently treated as a case of Sudden Infant Death Syndrome, otherwise known as 'cot death'. Williams had found some evidence of trauma in the form of bruises on the body but attributed these to harm caused during attempts to resuscitate the child.

Almost a year later, on 29 November 1997, Sally gave birth to a second son, who she named Harry. Despite being three weeks premature, Harry was a normal, healthy baby. To help her cope with the loss of her first child, Sally Clark received counselling and advice as part of a special programme for parents who had suffered a cot death.

On the evening of 26 January 1998 both parents were at home with Harry when Stephen Clark left the living room to prepare a night feed. A few minutes after he left the room Sally called out to him to explain that Harry had suddenly become unwell.

An ambulance was called at 9.27 p.m. When it arrived, the

team found Stephen Clark kneeling beside his unconscious son on the floor. The child was taken to hospital and further attempts to resuscitate him were made, but they were unsuccessful. Harry Clark was pronounced dead at 10.41 p.m.

The autopsy was once again performed by Dr Alan Williams, but this time he found evidence that Harry had been violently shaken on several occasions over the course of several days and that this was the most likely cause of his death. This conclusion made him reconsider his findings in the death of Christopher Clark and, having re-examined what remained of the original evidence (Christopher himself had been cremated), he concluded that his death had also been unnatural and may have been the result of deliberate smothering.

As a result of Williams' findings, Sally and Stephen Clark were arrested on suspicion of murdering their two children. The case against Stephen was quickly dropped, but police continued to pursue Sally. Having initially given a lengthy account of the circumstances leading up to the death of both children, in which she strenuously denied causing either any deliberate harm, she refused to answer any further questions at subsequent interviews, on the advice of her solicitors.

The case against her seemed strong, at least in the eyes of the CPS: both children had died at around the same age, both had been alone with their mother at the time they fell unconscious and both had just been fed before becoming unwell. Charged with two counts of murder, Sally Clark – by now pregnant with her third child – went on trial at Chester Crown Court in October 1999.

The strongest evidence for the prosecution was provided by Sir Roy Meadows, former professor of paediatrics at the University of Leeds, who appeared as an expert witness. He testified that the chance of two children from an affluent, non-smoking family suffering a single cot death was one in

8,543. On this basis, he concluded that the probability of the same family suffering two cot deaths was therefore 8,543 x 8,543 – one in 73 million.

Meadows told the jury this made the two cot deaths about as likely as backing an 80–1 outsider in the Grand National four years running and winning each and every time. He further argued that a double cot death would occur in the UK only once every hundred years. 'You have to say two very unlikely events have happened and, together, it's very, very, very unlikely. One sudden infant death in a family is a tragedy, two is suspicious and three is murder unless proven otherwise.'

Sally Clark was found guilty by a 10–2 majority and given a mandatory sentence of life imprisonment.

Professor Meadows had given his expert opinion without any objection from the defence and it clearly had a powerful influence on the members of the jury, who found Sally Clark guilty, but the problem was that his facts were completely and utterly incorrect.

The trial judge had clearly suspected something was wrong and had tried to steer the jury away from this part of the evidence: 'I should, I think, members of the jury, just sound a word of caution about statistics. However compelling you may find them to be, we do not convict people in these courts on statistics. It would be a terrible day if that were so. If there is one SIDS death in a family, it does not mean that there cannot be another one in the same family.' Meadows had treated each cot death as a completely independent event, the first unrelated to the other.

If you have a coin and toss it into the air, there is a 50–50 chance of it coming down heads; the same for tails. The probability of it being one or the other is therefore ½. Because each toss of a coin is an independent discrete event, the probability of tossing a coin twice and it coming down heads on both occasions is ½ x ½, which equals ¼. This was the logic Professor Meadows applied to his calculations.

However, with the cause of SIDS still largely unknown, the possibility exists that genetic and environmental factors may play a large part. On this basis, it's actually more, not less, likely that a family who has already suffered a cot death might go on to suffer the same tragedy once more. (Research would later suggest that a family suffering a cot death is between five and ten times more likely to experience it again.)

In October 2001 the Royal Statistical Society issued a public statement expressing its concern at the 'misuse of statistics in the courts' and noted that there was 'no statistical basis' for the figure of one in 73 million.

Adding to the misinformation provided by Meadows, the court had also committed the statistical error known as the Prosecutor's Fallacy. Many press reports of the trial reported that the 'one in 73 million' figure was the probability that Clark was innocent. However, even if the 'one in 73 million' figure were valid, this should not have been interpreted as the probability of Clark's innocence. In order to calculate the probability of her innocence, the jury would have needed to weigh up the relative likelihood of the two competing explanations for the children's deaths. Although double SIDS is very rare, double infant murder is likely to be rarer still, so the probability of Clark's innocence was in fact quite high. In reality, the odds for double SIDS to double homicide are between 4.5–1 and 9–1.

When a cot-death mother is accused of murder, the prosecution sometimes employs a tactic such as this: if the parents are affluent, in a stable relationship and non-smoking, the prosecution will claim that the chances of the death being natural are greatly reduced, so, by implication, the chances of the death being a homicide are greatly increased. But this implication is totally false, because the very same factors which make a family low risk for cot death also make it low risk for murder.

The errors made by Professor Meadows were serious enough, but they paled alongside those which it emerged had been made by the pathologist, Dr Alan Williams. Microbiological test results came to light showing the presence of Staphylococcus aureus at eight sites on Harry Clark's body, including cerebrospinal fluid, raising the possibility that Harry had died from natural causes.

This evidence had been known to Williams since February 1998 but was not shared with other medical witnesses, police or lawyers. It had not been 'peer-reviewed'. Williams had decided the report was not relevant and was able to 'self-certify' it as such, preventing it from being disclosed to the defence in the case.

After three years in prison and two appeals, Sally Clark was released in January 2003.

Her release prompted the Attorney General to order a review of hundreds of other cases. Two other women convicted of murdering their children, Donna Anthony and Angela Cannings, had their convictions overturned and were released from prison. Trupti Patel, who had been accused of murdering her three children, was acquitted in June 2003. In each case, Roy Meadows had testified on the unlikelihood of multiple cot deaths in a single family.

Williams was not the only Home Office-registered forensic pathologist whose competence was being questioned. In 2002 Kenneth Fraser appeared before an Old Bailey jury accused of murdering his partner, Mary Anne Moore, the previous year. Her body had been discovered at the foot of a flight of stairs.

In spite of the glowing reference I had given him at the Kent officers' mess many years previously, Home Office-registered pathologist Dr Michael Heath gave evidence that her fatal injury was caused not by a fall but by impact with a sharp-edged surface or object.

He later conceded that it was unreasonable to exclude the possibility of an accidental death, and he unreasonably failed to give any weight to the opinions of his experienced colleagues from the defence teams at the trial. He also agreed that his post-mortem examination of Miss Moore was 'so inadequate it significantly compromised the ability of the prosecution and the defence to explore matters material to the cause of death'.

Fraser was acquitted. A few months later Heath was called in to examine Jacqueline Tindsley, whose body was found in her bed at home in Lowestoft, Suffolk, in March 2002. Her partner, Stephen Puaca, was jailed at Norwich Crown Court later that year for her murder, following Dr Heath's evidence that she had been asphyxiated.

The conviction was later quashed by the Court of Appeal, as the judges heard from seven pathologists that there was no pathological evidence to support his view. They believed Miss Tindsley had died from an epileptic seizure after a drug overdose. Dr Heath had relied upon post-mortem observations in the Tindsley case that were not supportive of his conclusions. He had also failed to carry out all necessary procedures and adequately document his examination. His statement had not undergone 'peer review' before submission to the police and CPS.

Then there was Paula Lannas, an experienced forensic pathologist and close colleague of Dr Heath who was criticized over a number of post-mortem examinations. One case that caused particular concern was a post-mortem examination carried out on a baby who it was believed had been murdered. So 'suspect' were Lannas's conclusions that even the prosecution had to tell the jury that it could not rely on her evidence.

The Home Office knew that it was ultimately responsible

for the pathologists who were on its register, but there was little in place to monitor the quality of their work or deal with deficiencies.

It had set up and subsidized a Professional Standards Committee, a sub-committee of the Home Office Pathology Advisory Board, however the committee had few, if any, teeth. It comprised mostly active forensic pathologists on the Home Office register, with a few histopathology professors from the Royal College of Pathology. They could, and did, review complaints against their colleagues (and occasionally themselves) but in fact had very little executive power. In the end, both Dr Lannas and Dr Heath took the Home Office to judicial review over their being struck from the Home Office register – at great cost to the public purse. Neither of their appeals was successful and they were removed from the register.

Dr Lannas, in particular, would do things in a certain way, and if people tried to argue that they should be done differently, she would simply say: 'Where does it say that?' And she was right. The Home Office didn't have any written guidelines in regard to the working practices of its pathologists, so there was no way to prove she was in the wrong.

I had around six civil servants working under me for my three-year project – which had been started up as a direct result of the Alan Williams cases – to try to install quality controls and nationally agreed standards into forensic pathology. I thought it would be a fairly straightforward task, but it turned out to be a nightmare.

One issue is that there are three factions of forensic pathologists on the register: those who are self-employed, those who work for universities in teaching hospitals and those who work for the NHS in hospitals. The members of the different groups have very different approaches to the work and, as a result, do not get on with one another one bit.

All three wanted the £16 million spent on a central unit, but each group insisted it had to be in their own backyard. The money was only available for three years and, at the end of that, it would revert back to the Treasury if it was not spent. I had interminable and tedious meetings with them all, but they were far more interested in settling old scores with their colleagues than in advancing the cause of forensic pathology in the UK. They were never going to agree with each other: I was herding cats.

By the time the project was finished and the money spent we had introduced a sustainable career path to encourage new forensic pathologist entrants to the Home Office register, funded several new forensic pathology training schools – including one at the original site of the Royal London Hospital (which had offered to name the school after me) – and introduced a comprehensive quality-control system which required forensic pathologists to have their statements peer-reviewed before submission to the police and courts. We published a code of practice and an effective disciplinary procedure, upgraded and built a number of forensic mortuaries and had beaten up the pathologists enough to force them to form professional group practices to provide cover for the whole of England and Wales. All this was supported by a bespoke, Internet-based IT system – which they steadfastly refused to use.

Back on the international stage, Dave Werrett, CEO of the FSS, was about to embark on a staggering course of double-dealing that was to sour his personal reputation with the FSS's biggest customer, the Metropolitan Police, permanently.

In spite of the political and strategic importance of Turkey, the United Kingdom had never had the lead on a project in this country. One particular project arose, its aim to assist Turkey in its preparation for EU membership by supporting the development of forensic capability. The objective was to underpin the democratic system, including respect for the rule of law, through the introduction of modern scientific methods into police investigative practices. Human rights and policing issues were high on the political agenda in Turkey and the EU had made such issues a top priority in their accession negotiations.

Clearly, the involvement of government departments and agencies in projects in Turkey was a key target for the FCO, which was keen to develop links there following the war in Iraq. The Prime Minister raised the issue of continued involvement in 'twinning' and with Turkey in particular in a letter to the Deputy Prime Minister in January 2003. The Metropolitan Police were particularly keen for involvement because of their ongoing problem with organized crime among the Turkish community in north London and their wider responsibility for counter-terrorism and its links with Turkey.

In 2002 Turkey had made an unsuccessful bid for EU funding for a project to 'strengthen the police forensic capacity'. The EU in Brussels asked the UK FSS to assist in the re-drafting of the project plan before re-submission by the Turkish side. This was duly done by my colleague from previous projects, Andy Williams, and the plan was then accepted for funding by Brussels.

In June 2003 Andy and I drafted the UK bid for the EU-funded project in Turkey, which involved a partnership between the FSS and the Metropolitan Police and was led for the UK by the Home Office, with myself as project leader. Dave Werrett agreed to support it.

Later, and unknown to the UK bid team, the German forensic laboratory approached Werrett to ask if the FSS could support its bid for the same project. Werrett agreed to do so but kept his decision from the original UK bid team. However, a copy of the UK proposal was requested by, and sent to, the head of marketing – the member of the FSS involved with the German bid.

While at an international conference of the European Association of Forensic Science in Turkey, Andy and I were informed, by the German bid team, of the involvement of the FSS in the German bid. Gary Pugh at the Metropolitan Police then contacted Werrett to express his concern at the development, and the FCO advised the FSS CEO that he could not be partners in two bids simultaneously.

After considerable FSS discussion and action, during which the UK bid was blocked, released and support for the German bid withdrawn and reinstated, Werrett decided to continue with both the UK bid, and to support the German bid for the same project. A bizarre scene ensued where Andy (representing the FSS), me (representing the Home Office) and Gary and a commander (representing the Metropolitan Police) watched the head of marketing and three FSS staff shepherded off to the German hotel by their bid team. The head of marketing later turned up at the UK team's hotel demanding to see the UK bid presentation, 'or I will tell Dave Werrett', he threatened. We weren't scared, and he didn't see the presentation.

Sadly, the German bid eventually won by virtue of 'additional distinct offerings' over and above those funded by the EU for the project. These 'extras' were thought to have consisted of additional, free DNA-sequencing equipment, which was openly expected by the Turkish side and considered 'not the done thing' by the UK FCO. The German bid team, on the other hand, was happy to oblige.

The FCO made a protest about the scoring system to the EU commission in Ankara but, naturally, nothing came of it. The Metropolitan Police were extremely unhappy with the outcome, believing that operational as well as strategic opportunities for UK policing had been lost by the behaviour of the FSS CEO. Without FSS support, it is unlikely that a German bid could have been mounted, let alone been successful. The unique advantage of the quality-management systems present in the FSS was effectively offered to both sides.

Once again the FSS had failed to see the bigger, strategic picture and to support the UK, preferring instead to provide a minor supporting role to the German bid in exchange for a little easy income. The relationship between the FSS and MPS suffered irrevocable damage as a result, and Werrett's integrity was called into question. I was later told that the Turkey project was a disaster for the FSS.

20

THERE WERE NO WITNESSES TO THE ATTACK THAT TOOK THE life of 10-year-old schoolboy Damilola Taylor, stabbed close to his home on the North Peckham Estate in November 2000. All that remained was a trail of blood, which a local carpenter followed from the pavement to the stairwell of a nearby housing block. There, on the second floor, he found Damilola slumped against a wall, bleeding heavily from a wound in his leg.

The boy was rushed to hospital but, despite the best efforts of the medical team, he could not be saved. Originally from Nigeria, he had been living in England for just three months. An autopsy would later conclude that the single stab wound to his leg, which had severed an artery, had been caused by a broken bottle.

The large amount of blood found at the crime scene meant there was a very good chance that some of it would have been transferred to the clothing or footwear of Damilola's killer or killers. If the investigation team could track down whoever was responsible, they would be counting on the skill of the team at the Forensic Science Service to find the evidence they needed to solve the case.

Suspicions quickly fell on a notorious local street gang known as the Younger Peckham Boys, which was active in the area and whose members were known to mug younger boys for their valuables and pocket money. Although Damilola had neither, detectives investigating the case theorized that he may have been targeted because of the distinctive silver Puffa jacket he was wearing.

Although the investigating team initially hit a wall of silence in the community, their persistence eventually paid off and several names were put forward. One girl claimed that one particular youth, 12-year-old Danny Preddie, a member of the street gang, had confessed to killing Damilola because of a fight after school the previous day.

At first it seemed that Danny Preddie had an alibi but, despite this, he remained of interest, due to his connections with the gang, and was ultimately one of eleven boys arrested by police in the early days of the inquiry. Clothes belonging to all suspects were seized, their homes searched and items made ready to send off to the FSS for examination. These items included a pair of trainers belonging to Danny Preddie and a black Giorgio sweatshirt belonging to his brother, Ricky Preddie.

At the time, the London laboratory of the FSS was struggling with a massive caseload. There were 200 ongoing murder inquiries and a further 200 rapes, many of which were high profile and required a significant proportion of available resources to be devoted to them. Although some of the FSS's other labs in the country had more capacity, detectives working on the Taylor case insisted that the exhibits stayed in London. They preferred it, as it had been part of the Metropolitan Police and had retained many of the same staff.

Although the FSS was now in a commercial market and operating alongside other providers, it remained the 'supplier of last resort' for the police, meaning it had no choice but to take on whatever work was sent its way. While the privately run labs could refuse cases if they were overworked or set up a contract that guaranteed them a minimum amount of work for a set time and then increase staff or purchase new equipment to meet increased demand, the FSS simply had to manage as best it could.

As the investigation into the Taylor murder continued, so the tide of exhibits making their way into the laboratory did the same. The training shoes and sweatshirt were both part of submission number 4, which arrived at the laboratory on 4 December and contained 50 items in total, all of which were marked urgent.

Submission 13 arrived on 22 December and contained a further 144 items, again all marked urgent. On 16 January 2001, submission 16 arrived. It contained 78 items, once more, all urgent. In total, Operation Seale, as the Taylor murder case came to be known, involved 40 separate submissions between 30 November and July 2001, with a total of 441 items.

Most of these items – 403 of the 441, to be precise – arrived between November and the end of January 2001 – within eight weeks of the murder and therefore during the busiest time of the murder inquiry, and all but three were marked urgent.

Four days after they arrived at the lab, the trainers and sweat-shirt from the Preddie brothers were examined by an assistant scientific officer who had joined the FSS in 1998, successfully completed all her training and was considered competent. In accordance with laboratory procedure, she donned a cap, gown, gloves and mask then cleaned her workbench with bleach to ensure it was sterile. She then examined the bag containing the pair of trainers to look for damage that could indicate possible contamination.

Satisfied that the bag was sound, she opened it and examined the trainers by eye, using lights and a magnifying lens where necessary to identify individual stains. She then carried out KM spot tests for blood on both shoes. Those on the left trainer were negative, and those on the right were mostly negative, except for two or three stains on the front, outer aspect of the shoe. (DNA analysis would later identify this blood as belonging to Danny Preddie himself.) She then performed a general KM test on the uppers, soles, sole ridges and shoelaces of both trainers. The results were negative.

The ASO wrote up her notes. Normally, the reporting officer in a case would look at every article examined by an ASO and compare it with the notes provided. This is necessary, as it is the RO who has to give evidence in court, not the ASO. For whatever reason, perhaps because of the huge workload and pressure to provide results as quickly as possible – not to mention the excellent track record of the ASO involved – the RO decided not to double-check any of the items she had examined.

The black Giorgio sweatshirt wasn't examined until 4 January, by a different ASO, who had only completed his training the previous year. Instructions sent with the item were for the sweatshirt to be taped for fibres and checked for any blood or DNA traces.

The ASO carried out a visual inspection – first under strip lights and then more powerful fibre-optic lights – and saw no blood. He then checked this with a general KM test to the outside and inside waistband, cuffs and neck, stretching out the material to reduce the height of the ridges and to ensure that the filter paper was making contact with any blood that may have been present. All his tests were negative.

Once again, the item was not checked by the RO in the case. Out of all the items of clothing submitted in the Taylor case, the FSS reported that none was found to possess any significant forensic evidence.

But the FSS was wrong. Remarkably, both ASOs had missed vital evidence – traces of Damilola Taylor's blood and fibres from his clothing – on the two exhibits, and because no senior RO had checked the results, no one had spotted the error.

The FSS wasn't the only one feeling the pressure – or making mistakes. The murder of the schoolboy had launched a wave of horror and revulsion across the nation to a degree that had not been seen since the murder of Stephen Lawrence some seven years earlier. The *Daily Mail* put up a reward of £50,000 for information leading to Damilola's killers.

Eager to secure a conviction, the police team investigating the murder could scarcely believe their luck when a 12-year-old girl came forward and told them she was willing to provide the vital evidence they needed: she had witnessed the murder from behind a parked car and could name the four boys responsible.

She was far from an ideal witness. Code-named Bromley, the girl was a compulsive truant and had a tendency towards 'dishonesty, violent and disruptive behaviour' when she was in the classroom. She was, however, all the police had, and a few weeks after she came forward the four youths were arrested and charged.

All other suspects, including Ricky and Danny Preddie, were immediately eliminated from the inquiry. The FSS was asked to check the shoes and clothing of the four suspects. No evidence linking them to the crime was found but the police decided to proceed on the basis of the evidence provided by their star witness alone. It was a risky move. Juries were fast becoming more familiar with forensic evidence and expected to see it produced in cases. This notion became even more pronounced in 2001, when Channel 5 broadcast the first episode of hit American drama *CSI*.

Not only did the show highlight the role of crime scene investigators and forensic science in modern law enforcement in the USA, it also had an influence in the UK. SOCOs in Norfolk began calling themselves CSIs, and increased applications to forensic science courses at universities and colleges became one of the fastest-growing educational trends.

The Damilola Taylor murder trial began in January 2002 and, with no forensic evidence to back up any element of the prosecution case, quickly descended into a debacle similar to that experienced by police during Colin Stagg's trial for the murder of Rachel Nickell. Bromley was instantly exposed as a greedy attention-seeker who had, in the words of defence QC Lady Ann Mallalieu, 'taken the police for a bunch of mugs'.

Her behaviour in court was disruptive and led to the judge throwing out her evidence and two of the four suspects being acquitted of all charges, even before the start of the defence case. The remaining two boys were cleared by the jury after it emerged that their mobile phones placed them two miles from the scene at the time of the killing.

Although I was still officially working for the FSS, because I was on secondment to the Home Office I was able to view the events going on there with a sense of detachment. It was not a pretty sight.

In 2002, soon after the Damilola Taylor trial had collapsed, the FSS stopped being the preferred supplier of forensic services for the Association of Chief Police Officers. Police forces were now freer than ever to take their work elsewhere.

The sudden boom in the outsourcing of forensic science spurred rapid growth in the private sector as new companies jumped at the chance to cash in. In many cases, former FSS workers, disillusioned by the way the organization was being run, left and set up private companies of their own in direct competition. Having once had a virtual monopoly of over 95 per cent of forensic science work in the UK, the FSS was soon struggling and rapidly losing market share.

Furthermore, increasing levels of competition among the other entrants into the market meant that prices were constantly being undercut. The value of the market itself was shrinking, falling from an estimated £200 million in 1990 to £120 million in 2001. Against this backdrop of a developing market in forensic science and the changing relationship between the police and the FSS, a review of the FSS was announced in July 2002, to be led by a consultant named Robert McFarland, a former Chief Executive with the BOC group, who had headed a number of government reviews of the criminal justice system. He was ably assisted by Andrea Orban, one of my former account managers, now with an MBA.

Published the following year, the McFarland review recommended that the FSS be run in partnership with the private sector, rather than being solely funded by the taxpayer alone, as this would increase the FSS's flexibility and relieve the government of responsibility while allowing it to begin to see a return on its investment. So, in addition to charging for its services, the FSS of the future would also be expected to operate at a profit. As a precursor to this, the FSS would first be transformed into a private company, wholly owned by the government – a GovCo – to ensure continuity.

Reactions to the review were mixed. In 2004 the House of Commons Science and Technology Committee began its own review, 'Forensic Science on Trial', reporting on its inquiry into the McFarland review's recommendations.

Prospect, the union representing workers in the forensic

science industry, was appalled at the announcement, making it clear that its members believed that it simply wasn't appropriate for forensic science in the UK to be run as a purely commercial business. This view was in complete contrast to the views of private companies offering forensic services to the police; they saw the proposals as creating a more level playing field for all those involved.

From my point of view, it seemed the writing was on the wall. Having experienced the services of both the FSS and its competitors both as a scientist and as a customer while working at Kent, there didn't seem to be any realistic prospect of success for such a scheme.

At Kent, I had managed to reduce costs considerably by bringing more and more forensic work in-house. Other forces were doing the same and, with pressure on budgets set only to increase, the amount of work left for the FSS and other forensic providers was bound to decline. The organization was already unwieldy and poorly managed, and I didn't see how it could possibly compete. I had no wish to be on the ship when it sank. I applied, and was accepted for, a permanent transfer to the Home Office.

When I was working at the Home Office, putting the finishing touches to the pathology project, I was able to become involved in wider forensic science issues. A few years previously, Gary Pugh had persuaded me to take over his role as secretary to the crime scene working group of the European Network of Forensic Science Institutes (ENFSI), which had been founded in 1995 by the directors of western European government forensic laboratories to share knowledge and experience. By this time, ENFSI had become the accepted authority for forensic science cooperation and development in Europe and a DNA working group had been instigated to establish international guidance on interpretation standardization and acceptance of DNA methodology – in Europe, at least.

My boss in the Home Office then thought it a good idea for me to run for the chairmanship, which was becoming vacant. The UK was just about to embark on its six-month presidency of the EU, and such positions were politically expedient. I romped home in the, admittedly unchallenged, contest, and became chairman. I appointed Andy Williams to replace me as secretary.

Between us, Andy and I cajoled the crime scene steering group into applying for European Commission funding for the 'development of international standards for crime scene investigation strategies and good practice in crime scene examination methods'. We were successful, and quite a lot of EU money was allocated to us.

The working group had already been persuaded by the European Accreditation body that a new international accreditation standard – ISO 17020 – was appropriate to apply to scenes of crime examinations, which had until then not been covered by any accreditation standards. We now had the money to pursue this goal.

The steering group organized a two-day international conference in Krakow, Poland, in October 2006, with the EU funding. The conference was hugely well received and it was unanimously agreed that ISO 17020 should be adopted throughout Europe.

The scenes of crime working group was to receive the first ever 'Working Group Award' for exemplary development in their field from the ENFSI board of directors for this piece of work. The steering group then started to plan the follow-up conference in Dubrovnik with the rest of the EU funding. Unfortunately, this was cancelled by the Home Office while my back was turned.

Back in the UK, the ACPO scenes of crime working group initially rejected the adoption of ISO 17020 as the standard in favour of ISO 17025, which they had already started. They were made to see the error of their ways, and today ISO 17020 is being sought throughout Europe and the UK for accrediting scenes of crime investigations.

The failure of the prosecution of the four youths for the murder of Damilola Taylor was a huge blow for the police but, whereas after the acquittal of Colin Stagg detectives on the case refused to believe they had made a mistake, it was clear that the youths who had been charged had had nothing whatsoever to do with the boy's murder. The real killer or killers were still at large.

In early 2004 a new senior investigating officer was appointed to the case, and one of his first decisions was to have the items seized from the original suspects re-examined. Rather than having the FSS take a second look, the team decided to take full advantage of the way the world of forensic science had changed and ask one of the independent providers, a company called Forensic Alliance Limited (FAL), to carry out the work instead.

Operating without any of the pressure the FSS had been under, and with the benefit of being able to view casework that had already been completed on the Damilola Taylor case, the FAL team quickly found a visible bloodstain belonging to the victim on the heel of a trainer used by Danny Preddie. Within the stain was a fibre. Months of testing, re-testing and checking followed before FAL was able to announce with sufficient certainty that the fibre was a match to the material in Damilola's trousers.

After more searching, a fibre matching a sweatshirt worn by Danny Preddie was found on Damilola's jacket, and two more fibres that could match the material in Damilola's trousers were found on one of Preddie's jumpers. When the scientists began work on exhibits linked to his brother Ricky, they found a spot of Damilola's blood on the cuff of the black Giorgio sweatshirt.

The new evidence uncovered by FAL was a huge triumph for the police working on the case and propelled the investigation forward to the extent that both Preddie brothers were arrested and charged with murder.

However, there was still a question to answer: how had the FSS missed those same bloodstains all those years earlier?

The man most wanting an answer to this question was Gary Pugh, the former colleague of mine from the FSS who had gone on to become director of forensic services for the Metropolitan Police. Having requested a copy of the original FSS file relating to the exhibits in the case, he called and asked me to visit him at his office in New Scotland Yard.

I was back on the outside of the FSS looking in – very much my preferred position – and because I therefore had a degree of independence yet was fully familiar with working practices within the London Laboratory of the FSS, I was the first person to spring to mind when Gary realized he needed a second opinion about what he had found in the file.

As I sat down, Gary opened the file to show me a Polaroid photograph of the back part of a single training shoe belonging to Danny Preddie and asked what I made of it. The mark on the heel was obvious. Not only did it look like a bloodstain, but the stain had also been ringed with yellow wax crayon, which usually indicated that it had produced a positive result during a KM test.

I told Gary that, in my opinion, the stain was so obvious there was simply no way anyone could have missed it – and the markings showed that they had not. Had the stain failed the KM test, it would have been marked with a cross, rather than a circle. Indeed, there were several crosses visible on other parts of the shoe showing negative results for those areas.

The reality was that forensic science still depended on the human factor. No matter how conclusive it could ultimately prove to be, any piece of evidence could be rendered completely worthless if the scientist assigned to the case simply failed to spot it or neglected to follow through.

When the FSS was informed about what had been found, it immediately demanded the return of the original file. Gary refused: the fact that the stain was present on the original lab photographs taken back in December 2000 was a vital part of the case. The original file was seized as evidence and the FSS was provided with a photocopy. They were furious.

Both the photograph and other elements of the file would now form exhibits in the criminal trial. Although hugely embarrassing to the FSS, the file was the perfect way to deflect any allegations that blood and other material had been added to the shoe at a later date in order to secure a conviction.

In an attempt to defend the FSS's position, Dave Werrett wrote a note to the Permanent Secretary of the Home Office suggesting that, contrary to the beliefs of both the police and the Crown Prosecution Service, the evidence found by FAL did not necessarily provide a strong link between the victim and the suspect.

By now I was firmly ensconced at the Home Office and, as the senior forensic scientist working there, the note was in time passed on to me. It left me utterly speechless. The CPS was in the process of building their case against the Preddie brothers, a case the Metropolitan Police had spent many years, and tens, if not hundreds, of thousands of pounds to hopefully bring to a conclusion, and the CEO of the FSS was potentially undermining the key evidence in order to save face.

By law, the note would have to be made available – disclosed

– to the solicitors acting for the brothers, and threatened to seriously muddy the waters. As a result, Werrett was made a hostile witness (one whose testimony contradicts the evidence of other witnesses) in the prosecution case and barred from future meetings at New Scotland Yard.

He then wrote a retraction, explaining that he wasn't an expert in the field, that the information in his first note had been hearsay and that he had written it without looking at any of the case notes produced by either the FSS or FAL.

The trial went ahead (Werrett was never called to testify) and, after a long series of court appearances, the brothers were cleared of murder but convicted of manslaughter. It was a result for everyone working on the case, but in the process of achieving it, the relationship between the FSS and its single biggest customer, the Metropolitan Police, had been permanently soured.

After the trial, the Home Office sponsored a wide-reaching investigation into exactly what had gone wrong at the FSS. Under questioning, the scientist who had examined the exhibits was unable to explain why she had failed to consult with her reporting officer, even though her training required her to do so, and to cut out a portion of the stain for further testing.

The system set-up within the FSS at the time assumed that the reporting officer would examine every article looked at by the assistant scientific officer and compare it to the notes they had produced, as it would be the RO who would be giving evidence in court. When questioned, the reporting officer admitted that, because of the enormous pressure he was under, he had temporarily abandoned 'best practice', which dictated that he should have double-checked the work.

The failure to find the evidence in the Damilola Taylor case was a PR disaster for the Forensic Science Service. Operating as it now was in the midst of a highly competitive market in which the police were able to spend their money wherever they thought they would get the best results, it was just the kind of setback it really didn't need.

The opposite was true for FAL, which received a massive publicity boost and which, within a year, had been bought by and merged with the other major independent player in the UK forensic science market – LGC.

Just a few months after the initial discovery of the bloodstains in the Damilola Taylor case, the House of Commons Science and Technology Committee announced its own review on the state of the market in forensic science and the prospects for the success of a public-private partnership.

I was invited to give oral evidence, appearing before the committee in December that same year. I took the opportunity to highlight the fragility of the market and the fact that the police themselves, not just other forensic providers, had to be considered part of the future competition. I followed this with written submissions the following year, pointing out that the pressure on police to cut their forensic science budgets was likely to lead to a reduction in the number of items being submitted to outside laboratories, further reducing the potential for any company – private or government-owned – to make a profit. The whole forensic science market would contract.

The committee's final report was sceptical of the whole process, especially given what it referred to as the 'government's poor track record at managing public-private partnerships', but, with FSS senior management supporting the move, it seemed that it was inevitable.

As the day of the move to becoming a GovCo drew closer, it occurred to me that, if the FSS was going to become what was essentially a private company, it could no longer perform the 'regulatory' role it had held until then. The FSS chief scientist had advised ministers on forensic science matters, introduced and maintained standards and investigated screw-ups. There was no provision for the future regulation of the forensic science market, independent or otherwise.

Having experienced first-hand the glaring errors that had led to the appalling delays in the Damilola Taylor case, it was clear that a new, independent watchdog needed to be appointed to watch over the whole forensic science market, carrying out a similar role to that performed by OfCom in the broadcasting and telecommunications market, or the Independent Police Complaints Commission with regard to the police.

My position at the Home Office allowed me to have direct contact with ministers. I sent a submission to the then Policing Minister, outlining the need for independent standard-setting and quality regulation in the forensic science market.

I received ministerial approval to carry out a consultation and put together a fuller proposal about exactly how such regulation should operate. As I began work, the Home Office announced its sponsorship of an investigation into forensic failings in the Damilola Taylor case, and I was once again asked to submit evidence, this time showing how a regulator would help prevent such mistakes from occurring in the future by ensuring minimum standards of best practice were adhered to.

The review made it clear that a Forensic Science Quality Regulator was desperately needed. It would take many months before an appointment could be made, so I agreed to perform the role in the meantime.

As it turned out, my services were needed almost immediately. As the interim regulator, I found myself dealing with a case that would shake the world of forensic science, cast doubt over thousands of convictions and acquittals throughout the UK and ultimately hasten the demise of the FSS itself.

21

THE BELIEF THAT THE FAILURE TO FIND DNA OR OTHER FORENSIC
evidence on an exhibit meant there was no DNA to be found
was no longer as solid as it had once been. The Damilola Taylor
case had demonstrated that it wasn't always about having the best
technology – some of it was down to the human element, and
staff who were under less pressure or more motivated had more
incentive to look that little bit harder.

Having succeeded where the FSS had failed on the Damilola
Taylor case, the Metropolitan Police decided to submit the
tapings from the Rachel Nickell case, which had failed to
produce a result using the super-sensitive LCN technique, to
Forensic Alliance to see if they could pull a rabbit out of the
hat once again.

Part of the motivation had been an anomaly in the results
the FSS had provided: not only had their tests not detected any
male DNA on the tapings, they had also failed to find any trace
of Rachel's own DNA, which should surely have been present,
in the form of loose skin cells.

An extract from the tapings was submitted to Cellmark (via

Forensic Alliance) in September 2004. Using standard profiling techniques based around the usual 28 cycles of PCR, the company at once produced two sets of DNA. One profile was female and soon found to belong to Rachel Nickell. The other set of DNA was male.

The quality of the sample was such that only a partial profile of the unknown male could be obtained. This was immediately compared to the profile for Colin Stagg. It was not a match. More than a decade after the murder, Stagg was no longer the number-one police suspect for the killing. The profile was then compared to other suspects in the case and a match was made: the DNA appeared to belong to Robert Napper.

The link to Napper was supported by red paint flakes found on samples from the victim's son that matched the paint on Napper's toolbox, and by a footwear mark at the scene which could have been made by one of his shoes.

Further testing and refinement of the male DNA sample was carried out in the months that followed, and it wasn't until February 2006 that multiple tests finally confirmed that the DNA belonged to none other than Robert Napper.

At that time, the FSS was still working hard to extol the benefits of its proprietary LCN DNA technique, but doubts about how effective it truly was were beginning to surface.

Why, for example, had the FSS failed to get a result from the Rachel Nickell tapings while using LCN, yet Forensic Alliance had managed to produce a full DNA profile using their standard SGM+ profiling methods? It was a question that needed answering, so Gary Pugh, as Director of Forensic Services for the Met Police, asked the FSS to conduct an investigation and report back to him. He requested that they confine their comments to their own internal investigation.

Asked to comment on their findings, the FSS immediately raised the possibility of contamination before going on to produce a more detailed report outlining the reasons behind its failure. It made surprising reading. The FSS claimed that the reason it had failed to get a result was because of the phenomenon known as PCR inhibition, something it was claimed had been entirely unknown to the FSS at the time the first set of analyses was conducted in 2001.

The FSS then casually explained that PCR inhibition had first been identified by their scientists in October 2003 and, as a result, their working methods had been changed in 2005 to prevent it from happening again. Furthermore, claimed the FSS, PCR inhibition was so widely unknown that, if the 2001 analysis had been carried out by Forensic Alliance, it, too, would have failed to produce a result.

Far from resolving anything, the FSS report simply generated more questions. Chief among them was this: if the FSS had found that LCN was not working in 2003 and had changed procedures in 2005 to rectify matters, what had happened in all the cases it had worked on in between using the faulty technique, cases that potentially stretched back all the way to 1999, when LCN was first introduced? And why had they not informed their customers – the police? In total, the technique had been used on more than 21,000 cases. How many times had the FSS told investigators that no DNA was present when in fact it was there to be found?

Pugh also had doubts about the explanations the FSS had given for the 2001 analysis failures. Eager to get to the bottom of it all, and, since the FSS had referred against advice to Forensic Alliance, he sent the FSS report to Forensic Alliance (which by that time had been taken over by LGC) and asked for its comments.

LGC reported back that not only was the problem of PCR inhibition well known in 2001, and therefore would not have been an issue had their company carried out the first set of tests, but that the issue was considered so important that a whole chapter of a key reference work had been devoted to it back in 1999. Besides, PCR inhibition was the very reason that the manufacturers of the DNA kits set the limit at 28 cycles in the first place, and warned of it in the instructions for use issued with their kits.

The FSS had made the erroneous presumption that there was not going to be sufficient DNA on the Rachel Nickell tapings to get a result using standard DNA analysis, so they had dived straight in with the more expensive LCN technique. If they had used standard DNA SGM+ analysis, as Forensic Alliance had done when they took over the case, they would have got a full DNA-profile result.

LCN was a proprietary system within the FSS and, the more successful it was, the more it would be requested by police forces. Initially, the technique was a last resort. If standard procedures failed to produce a result, you could try LCN with the material you had left. The failure rate was still high, though, with only 6 per cent of cases producing a full profile. Had commercial pressure pushed the FSS into using the technique?

From the reports Gary Pugh had received it was clear that not only was the FSS attempting to explain away its original errors, but it was also sweeping under the carpet thousands of analyses it had conducted that had produced negative results which, had the technique been properly applied, might have been successful.

Many of these cases had already been through the court system and, though doubts about the reliability of the DNA evidence might have been expressed by counsel for the defence in some cases, none had been high profile enough to bring the matter to wider attention. All that was about to change.

★

In August 1998 members of the Real IRA, a splinter group formed by former members of the Provisional IRA opposed to the Good Friday Peace Agreement, set in motion a plan to attack the court-house building in Omagh, County Tyrone, using a car bomb.

Just after noon on the thirteenth a red Vauxhall Cavalier had been stolen in the Republic of Ireland and driven across the border, packed with 230 kilos of fertilizer-based explosive. Unable to find a parking spot outside their intended target, the two male occupants parked some 400 metres away, in the midst of a crowded shopping area, armed the device and made good their escape.

That afternoon, a series of warnings was issued to various arms of the media, all of them citing the court-house as the target. The Royal Ulster Constabulary began evacuating the area and was still doing so when, 40 minutes after the initial warning, the device exploded.

Confusion about the actual target meant the wrong area was being cleared and that surrounding the red Cavalier was full of people. Twenty-one people in the immediate vicinity died instantly, and eight more on the way to hospital. The victims included six children, six teenagers and a pregnant woman. More than 300 people were injured by the blast.

The first arrests in connection with the bombing were made the following month, but all those detained were released without charge. As the years went by, more arrests followed, as did more acquittals. By 2006 only one man had been convicted of involvement with the explosion: builder and publican Colm Murphy, though his criminal conviction was to be overturned on appeal and he would subsequently be acquitted in a retrial. Then, that summer, Murphy's nephew, Sean Hoey, was arrested and charged with 29 counts of murder, along with numerous other terrorism charges. He had been linked to bomb fragments by LCN DNA analysis conducted by the FSS. Doubts about the veracity of the technique emerged early on, when a sample taken from a car bomb in Lisburn, Co. Antrim, was wrongly linked to a 14-year-old schoolboy in Nottingham.

As the Hoey trial began, the judge at Belfast Crown Court, Mr Justice Weir, learned that the LCN process was admissible as evidence in only two other countries in the world, New Zealand and the Netherlands. With such high stakes, the judge wanted to find out more. Dr Peter Gill, one of the principal architects behind the development of LCN DNA for the FSS, was called to give evidence.

Never having been a reporting officer, Dr Gill had little experience in the witness box and was clearly uncomfortable throughout. During his cross-examination, the topic soon turned to that of validation: had LCN been independently tested, verified and accredited the way that other forms of DNA testing had been?

He was ultimately forced to admit that the only validation of the technique had been carried out in-house by the FSS itself.

In early November 2006, soon after the Hoey trial had begun, ministers were informed of the potential problems with LCN DNA by Chief Constable Tony Lake, the ACPO Forensic Portfolio lead, who was asked to conduct a review on behalf of the Home Office. In my new role as forensic science regulator, and working hand in hand with ACPO, the Crown Prosecution Service and others, we launched an investigation.

At first, the FSS intended to investigate the problem internally and refused to cooperate. It did so only after serious ministerial intervention.

My first priority was to ensure that the processes of all forensic science providers which undertook analysis of very low levels of DNA were not susceptible to inhibition. As forensic science regulator, along with the Operation Cube (as the review was known) scientific adviser, Dr Steven Rand, I visited key forensic suppliers in order to assess the LCN DNA – or LTDNA – techniques that were being employed by them. We produced a report in August 2007 which outlined that the (FSS) LCN updated procedures that were now being used by the FSS for re-analysis and ongoing LCN casework were appropriate to analyse DNA in samples where inhibition had potentially occurred. The report confirmed that LTDNA techniques used by other forensic science suppliers had always been and remained fit for purpose. We also reported that, as suspected, the FSS had 'verified' their (FSS) LCN technique internally but not validated it against any internationally or nationally recognized technical standard.

Our report was eagerly awaited. The judge in the Hoey trial chose to delay giving his summing-up until he had a chance to read it. The judge expressed 'concern about the present state of the validation of the science and methodology'. Soon afterwards, Hoey was acquitted of all the charges against him.

From the outset of Operation Cube the CPS responded in dramatic form, ordering the re-examination of cases involving LCN currently going through the courts, and the Association of Chief Police Officers also announced it was suspending its use of LCN DNA while the review was undertaken.

As a result of our investigation it became clear that, over a five-year period, the Forensic Science Service, using LCN, had failed to detect any DNA in samples from 2,500 cases including murders, rapes and serious assaults. It was also clear that in many cases DNA samples could have been found and potentially matched to suspects, had the FSS used different techniques, such as those that were being used by other, privately run, laboratories.

Chief Constable Tony Lake told the media that the cases in

which the FSS had failed to analyse traces of DNA involved the most serious crimes. 'This is about not getting results when it might be expected that there was DNA, rather than getting a result that was wrong. This type of DNA analysis of tiny amounts of DNA is carried out normally in the most serious crimes. We were not best pleased. We were not impressed. We rely on our forensic providers to have the highest standards.'

Lake wrote to all chief constables, asking them to examine their files for any cases between 2000 and 2005 that had not resulted in a conviction in which samples for LCN analysis had been sent to the FSS and returned negative when a positive result might have been expected. In all, 4,841 samples were identified and re-analysed, providing new information in 342 cases nationwide, 15 of which were described by police sources as 'significant'.

However, it wasn't always possible to re-analyse old cases. In some instances there had been so little DNA available in the first place it had been used up in the initial testing.

LGC offered to help, but the FSS refused. This wasn't surprising. It was clear that there were going to be many instances when new evidence was found as a result of the re-analyses, and the FSS didn't want another company making capital out of its mistakes. Again.

The FSS absorbed all the costs of the re-analysis operation itself, as demanded by ACPO and the Home Office. All the tests were done using LCN, but this time with the added dilution step, which by this time had been fully validated. The process used up valuable equipment and manpower, as well as expensive reagents and other consumables. With reported losses at the time of £2 million per month, there can be little doubt that this huge expense added greatly to the financial woes of the organization.

The following year, 2008, a government-commissioned

review into the effectiveness of LCN DNA led by Professor Brian Caddy gave the technique the all-clear, in part at least because of the changes in operational procedures the FSS had already made by the time the review started. The brief suspension of the technique was reversed and it returned to use.

In essence, the FSS had already identified all the potential problems back in 2003 and put measures in place to prevent them from occurring in 2005. The key issue was that it had not told anybody about it at any stage.

However, because the technique had failed on so many occasions, despite the fact that DNA was present, the FSS acquired a bad reputation, and this contributed to its downfall.

As if the problems with LCN DNA were not enough, the FSS also had another problem on its hands.

In October 2006 the company announced that it had developed a new technique to enhance the number of matches it achieved on the National DNA Database (NDNAD) – its 'hit rate'. This time, the system was computer-based, and it was claimed that it could interpret previously unintelligible, mixed DNA samples.

Routinely, the NDNAD would compare crime scene profiles against the database of individuals – scene-to-individual searching. But scene-to-scene searching could also throw up linked offences across the country, and aliases identified by individual-to-individual searches, if needed. Sometimes a profile match would show that the wrong person was in jail. Sometimes there would be a match for a murder: the profile of a suspect would match a profile from a crime stain. If a match was not obtained, the undetected crime stain profile would remain on the database and be routinely checked against all new profiles added.

The average discriminating power of a 'full' profile is one in a thousand million. Crime stains, however, do not always yield a full profile, with information at fewer than normal allele sites. These 'partial' profiles have a lower discriminating power and may match a number of individuals on the database.

Partial profiles with a discriminating power of at least one in a million could be loaded on to the NDNAD and searched against all records held on a daily basis. By this time the database contained about 1.2 million DNA profiles of people convicted or awaiting trial, and about another 108,000 from stains found at crime scenes. This had given it the kind of critical mass which meant that there was now a 40 per cent chance of a newly arrived stain being matched to an individual.

It had also reached the level at which 'coincidental hits' could be expected. Common sense dictates that if the 'probability of a chance DNA match is one in a million' then, once a database gets to hold a million DNA samples, the likelihood is high that one of them will match by coincidence. Coincidental matches in very large databases are referred to as 'adventitious hits'.

In the space of just five years the database had discovered nearly 125,000 matches − 112,000 of them between suspects and crime stains, the rest showing either that another crime had been committed by the same person − as yet unidentified − or that someone on the database had been using an alias. However, chief constables and government ministers were all too aware of scientific developments which raise the spectre of Big Brother and excite the civil-liberties lobby: they're aware that mistakes can be made, samples can be contaminated, and that the courts usually want more than just DNA evidence to convict someone.

Yet their high hopes for DNA were clearly illustrated by the fact that the government, through the DNA Expansion Programme, allocated an extra £202 million to increasing the number of scenes of crime officers trained in DNA-gathering techniques and to expanding the National DNA Database from its current size to more than 3 million – the generally estimated total of the criminally active population. The investment amounted to around £3 for every man, woman and child in the country.

This threefold expansion of the database happened mainly as a result of a change in police habits: previously, they took DNA samples only from suspects in the more serious crimes, but they had increasingly exercised their legal powers to take them from almost everyone they arrested. The scientific techniques were becoming so powerful that they didn't know whether they might clear up a string of burglaries from a cigarette end or even a flake of dandruff left at the scene.

Another surge in numbers arrived with the abolition of the practice of taking samples off the database when people were acquitted by the court or no action was taken against them after their arrest. This change was the most controversial clause of the Criminal Justice and Police Act (2001), which became law three days after the general election was called – but not before some fierce arguments both in and outside Parliament.

Known as DNABoost, the new FSS technique was hugely appealing. It claimed to maximize the benefits of LCN and to help sort out some of the issues such a high level of sensitivity created. In cases where mixed DNA profiles were obtained, DNABoost was said to be able to help clarify their interpretation. In all, it was stated by the FSS that DNABoost would increase searchable DNA profiles by up to 15 per cent. It wasn't so much a new technique for DNA analysis but rather a software development which could help to interpret mixed profiles, examining their enhanced individual components and then cross-checking these against a modified DNA database.

The FSS chose to go directly to regional CPS Chief Crown Prosecutors in order to 'pilot' the technique on live cases in West Yorkshire, South Yorkshire, Northumbria and Humberside police forces. Two hundred and ninety-one DNA profiles had already been processed when the regional CCPs looked to the national policy lead at CPS headquarters for guidance, who in turn looked to me.

In my role as forensic science regulator I made an assessment of the technique from what I was shown and told by the FSS, and advised the CPS of my concerns. The Director of Public Prosecutions immediately wrote to Dave Werrett at the FSS, withdrawing CPS support from the pilot projects and suspending the use of DNABoost in 'live' criminal cases.

What had struck me as odd about DNABoost was that it appeared that it would only be able to work – and could only have been developed – if the FSS had access to a copy of the National DNA Database and had modified it – and I didn't think they were supposed to do this. Clearly, the DPP had agreed.

I was invited to a meeting at CPS HQ on 1 December 2006, along with Dave Werrett and a few FSS research scientists who were there to detail the workings of DNABoost. They explained that approximately ten to fifteen thousand DNA results per year, in cases with no suspects, end up as partial, degraded or mixed profiles. DNABoost, they went on, enabled a search on a separate database of suspect DNA profiles and would facilitate an association being made in the first place – it is an 'intelligence tool'. Suspect samples are 're-coded' to allow for more discriminatory searching by DNABoost.

However, Werrett was very cagey about exactly how 're-coding' was done, claiming that the process was FSS intellectual property and 'secret'. It was claimed that DNABoost was a software package that allowed more discriminatory searches

– but the original individuals' DNA data would have to be manipulated (re-coded) in some way to allow for this greater discrimination. When pressed, Werrett did say that the FSS would disclose the exact method under examination in court if required to do so.

The CPS was not satisfied or happy with this explanation, pointing out that it would lead to problems if the evidence was not contested for several years and was then shown to be flawed – rather surprisingly like LCN, in fact. It seemed that the FSS had learned nothing from the LCN disaster.

I was not alone in my concern that the 'Boost' nominal and scene DNA databases appeared to be privately held, searchable databases of individuals' DNA under no governance or ethical supervision. It remained to be seen whether these databases were maintained and searched legally, ethically or with scientific rigour. It was questionable whether it was legal or ethical for a 'private' supplier of DNA results to manipulate and retain data from individuals in order to use this data for research and subsequent commercial advantage.

Those not of the FSS concluded that DNABoost, if subsequently deemed to be reliable and efficacious, would serve justice better if applied to all samples on the NDNAD. This could be easily effected, while retaining the FSS commercial interest, by virtue of licences, which the FSS could sell to competitors in the forensic DNA market.

A follow-up meeting, chaired by the CPS, was arranged, which concluded that the correct procedure would be for the FSS to apply immediately to the National DNA Database Strategy Board (of which Gary Pugh was chair) for retrospective permission to use the DNA data in the FSS's possession for research and commercial purposes.

The FSS applied; the NDNAD Strategy Board refused permission. Almost as quickly as it was born, DNABoost was suspended and yet another nail had been hammered into the coffin of the FSS.

The post of forensic science quality regulator that I was standing in for still had to be formally filled through the open public appointments process, but it was generally expected that I would apply for and obtain the position. However, I had had enough of the petty in-fighting between the forensic pathologists and of dealing with the FSS, so I took a year's 'career break' instead.

The Home Office sent me the application forms, but this time I was determined not to apply. I saw the post as a poisoned chalice and I also strongly believed that, for the sake of the independence of the regulator, the post should go to a person who favoured neither the police 'customer' nor the forensic science providers – perhaps a retired judge. In the end, a civil servant who was temporarily replacing me in the Home Office chose the expedience of an ex-policeman.

When I returned to my desk in Marsham Street, in May 2008, the world had changed again. I was now working as adviser to the newly appointed regulator. His first question to me was: why had I not applied for the post? What was he missing? I chose to let him find out for himself.

In October 2008 the FSS launched a business-transformation programme aimed at making the organization into a profitable enterprise and providing value for both its shareholders and the criminal justice system. The transformation was supported by a £50 million government grant.

As it got under way, it became clear that Dave Werrett was not the man to take the organization into the future. Despite having presided over its woeful recent history, including independent inquiries into the Damilola Taylor and Rachel

Nickell fiascos, Werrett received a payoff of £875,000. Details of the payment even made the *News of the World*, which described Werrett as one of three 'bungling bosses' of the FSS and quoted a government source that said, 'These top staff have walked away with a bumper bonus in an organization whose ineptitude let killers and rapist scum escape justice.'

A new management team took over, with the chairman initially acting also as chief executive, but, as far as the FSS was concerned, it was too little, too late. All the damage had already been done. With its reputation in tatters and so much of the work it might have been able to do itself now being carried out by other providers, not to mention the shrinking size of the market itself, it seemed the writing was on the wall.

Once the decision had been made to put a price on justice and for the UK to become the only country in the world where government-owned forensic science services are required to make a profit, it was only a matter of time before the inevitable occurred.

Personally, I had had enough of the Home Office and forensic science. It had changed beyond recognition from when I had started in 1978 and was no longer a career I would have embarked on if starting out now. So, in January 2009, I resigned.

And, then out of the blue, came a bombshell. On 14 December 2010 the Home Office announced that the FSS would be closed down in March 2012:

> Despite this intervention [£50 million government grant] and the commitment of the current management team, the current challenging forensics market has put the FSS back into serious financial difficulty. FSS is currently making operating losses of around £2 million per month. Its cash is due to run out as early

as January next year. It is vital that we take clear and decisive action to sort this out. The police have advised us that their spend on external forensic suppliers will continue to fall over the next few years, as forces seek to maximize efficiencies in this area. Her Majesty's Inspectorate of Constabulary concurs with this assessment.

We have therefore decided to support the wind-down of FSS, transferring or selling off as much of its operations as possible. We will work with FSS management and staff, the Association of Chief Police Officers, and other suppliers to ensure an orderly transition, but our firm ambition is that there will be no continuing state interest in a forensics provider by March 2012.

22

THE FACT THAT SHE WAS A FEMALE SERIAL KILLER MADE HER enough of a rarity; it was the fact that she was also responsible for a wide range of far more petty crimes that had detectives across Europe scratching their heads in bewilderment.

The killer first came to public attention in April 2007, when 22-year-old police officer Michelle Kiesewetter was shot dead in her patrol car in the city of Heilbronn in the south of Germany. Kiesewetter was sitting alongside her patrol partner on a lunch break when the killer approached and shot her in the head at point-blank range. Her partner was also shot in the head, falling into a coma for several weeks before awaking with no memory of the attack. The guns and handcuffs of both officers were stolen.

The brutal execution-style murder made headlines across Germany and Heilbronn police made tracking down the culprit their number-one priority. Traces of DNA found at the scene showed the killer was female, but a search of the German DNA database did not reveal her identity.

Police then searched through previous cases and were

startled at what they unearthed. The same DNA had been found on a cup in the home of a 62-year-old woman killed in another German town back in 1993. The DNA had also been discovered on a kitchen drawer following the murder of a 61-year-old man in the town of Freiburg.

The following January the bodies of three Georgians were found in a car in Heppenheim. When tests were conducted, the same female DNA was detected at the scene. It turned up again in October in the car of a murdered nurse.

No one had ever seen her and no CCTV camera had ever captured her image, no matter how extensive the coverage in the area where she had chosen to strike. The German pressed dubbed her 'The Woman without a Face' and 'The Phantom of Heilbronn'.

When the German police widened their search parameters, they found traces of the same mystery DNA in France and Austria. No murders had been committed in either country: instead, the Phantom seemed to have been involved in a range of minor crimes. Her DNA was on a toy pistol used to rob a Vietnamese jeweller's shop in Arbois, France; at the scene of a burglary of an optometrist in Austria; and at the scenes of dozens of car and motorcycle thefts in both Austria and Germany. In all, the Phantom was linked to more than 40 separate crimes.

It wasn't until March 2009, by which time Heilbronn police alone had racked up more than 16,000 hours' overtime, that the mystery was finally solved. Investigators examining the charred body of an asylum seeker in France were astounded when DNA analysis showed that the body was that of their elusive phantom killer. The problem was, the body was male, and the DNA results they got back were very much female.

Further investigation then uncovered the cause of the

mysterious results – not a phantom killer but the swabs the police forces had been using. All produced by the same Eastern European factory, the swabs had been contaminated with the DNA of one of the workers as they were being manufactured, and increasingly sensitive profiling techniques had continually produced hits, all linked back to this one worker. The case highlighted the issues and difficulties that all forensic labs now face, as LGC itself was about to discover. It was a deeply embarrassing blunder for the German police, but also exactly the sort of situation that was likely to become increasingly common (as DNA analysis continues to advance) unless appropriately rigorous quality standards were introduced.

In the months that followed the announcement of its closure, the Science and Technology Committee conducted its seventh report into the Forensic Science Service. Published in July 2011, the report stated, 'The common view appears to be that neither the current nor previous government handled the FSS's situation particularly well.'

During the course of its investigations, it emerged that neither the Director of Public Prosecutions nor the head of the CPS had been directly consulted. The Criminal Cases Review Commission had not been informed in advance at all, despite the obvious impact that closing the FSS was about to have on the criminal justice system.

Alternatives to closure had, apparently, been considered but, because of the FSS's status as a GovCo, it was subject to the Companies Act and had to comply with company law. This left only three potential options.

The first was to allow the FSS to go into administration – obviously a non-starter. The second was to restructure the FSS through further investment – something the government was not prepared to do, due to the expense and the risk that, ultimately, it might not be able to sell the FSS for anything like the amount of money it had invested in it. The third, and chosen, option was to wind the company down.

An additional option, which was never fully considered, however, was to split the FSS, separating the profit-making and non-profit-making functions in order to retain the ability to carry out research and deal with complex cases or those that involved national security.

When I started my career in forensic science, the general philosophy towards physical evidence was straightforward: if it was too small to see, it was too small to be worth bothering with. In those days, there was no value in finding a tiny trace of blood, as it would not be possible to identify even if it was human, let alone ascertain who it came from. The situation today is very different, with the advent of LCN DNA allowing full profiles to be obtained from microscopic traces of blood and body tissue. Exhibits that were once considered useless can now be subjected to a whole new level of scrutiny, which, in many cases, yield positive results.

Most recently, this was demonstrated in the conviction of two men for the murder of London teenager Stephen Lawrence. It stemmed from a microscopic examination by LGC of clothing worn by one of the accused.

This conviction, along with a number of others, has been portrayed as clear evidence of the benefits of privatization but, in reality, the picture is not quite so clear-cut. Much of the work in the Lawrence case, for example, was attributed to the huge advances that had been made in forensic science generally in the decades since the murder.

LGC did not pioneer anything new in order to crack the case, and the fact that the company had missed the same bloodstain during an earlier review of the case was not widely publicized.

★

The possibility nowadays of obtaining highly specific DNA results from microscopic traces has led to much greater anti-contamination measures and more precise note-taking than was ever the case in the days before DNA. Nevertheless, cross-contamination can rarely be ruled out absolutely, even with today's carefully recorded case files and scene notes, due to the uncontrolled nature of the original crime scene and early investigative process. This was particularly the case in the Stephen Lawrence investigation, where early investigators were heavily criticized for incompetence. Mistakes do and will continue to happen.

In 2010 taxi driver David Butler was arrested for murder. His DNA had been found under the fingernails of a woman who had been killed in 2005 and whose case was being reviewed. His DNA sample was on record because it had been detected on a cigarette butt following a burglary at his mother's home some years earlier. The sample was only a partial match, of poor quality, and experts at the time said they could neither say that he was guilty nor rule him out. Nevertheless, he was charged and remanded into custody.

It emerged that Butler has a rare skin condition, which means he sheds flakes of skin, leaving behind much larger traces of DNA than the average person. Thanks to his work as a taxi driver, it was possible for his DNA to be transferred from his taxi via money, or another person, on to the murder victim. The victim was also wearing a glitter nail polish which proved particularly attractive to dirt – and DNA.

The case went to trial and Mr Butler was acquitted and finally released after spending eight months in prison. The real killer has never been caught and the case remains open.

Cases such as this call into question whether there are larger issues with the forensic landscape in the UK. The move to the open market means that every test now has a financial im-plication. As a result, it is the police rather than the forensic scientists who take the lead when it comes to deciding which tests to perform. And sometimes the people making the decisions lack the necessary specialist knowledge to enable them to get the best result.

Some solicitors complain that DNA has made the police lazy: rather than taking a closer look at all the evidence, they pin everything on the DNA results. With so much authority and certainty being placed in the power of DNA, along with the fact that profiles can now be obtained from microscopic samples, the potential for error can only grow larger.

In October 2011 officers from British Transport Police re-sponded to a street fight involving a number of youths who had gathered near Exeter. Among the allegations made were that the youths had been spitting and, in order to ascertain which of the young men had been responsible, swabs of the spittle were taken, along with samples from all the youths present, one of whom was 19-year-old Adam Scott from Truro.

The swabs were sent to the LGC laboratory at Teddington for DNA analysis, arriving at the lab on the sixth.

The following day, the LGC's other laboratory in Risley received six swabs and a set of clothing taken during the forensic medical examination of a woman who had been attacked and raped in Blackley, Manchester, at the beginning of the month. An initial examination showed that semen was present on each of the swabs, so these samples were separated from other cellular material and sent to the Teddington lab to be analysed for DNA.

The low and high vaginal swabs produced DNA from semen that was quickly identified as belonging to the victim's boyfriend. Two of the swabs produced a mixed profile, as they contained semen from the boyfriend and an unknown male, assumed to be the rapist. This second profile was loaded into the National DNA Database and produced a hit: Adam Scott. The match was not a full profile. Instead, 17 of the 20 alleles corresponded to produce what is known as a partial profile, but this was considered more than adequate to proceed.

On 22 October an LGC scientist forwarded a summary of her findings to the Greater Manchester Police, who were in charge of the rape investigation. Her report concluded, 'It is estimated that the chance of obtaining matching DNA components if the DNA came from someone else unrelated to Adam Scott is approximately one in 1 billion. In my opinion, the DNA matching that of Adam Scott has most likely originated from semen.'

The subsequent witness statement affirmed, 'In my opinion, the scientific findings in relation to [the victim's] vulval swab provide strong scientific support for the view that Adam Scott had sexual intercourse with [the victim] rather than that he did not.'

Scott was arrested on suspicion of rape that same afternoon and driven from Devon to Manchester in order to be questioned. During interviews under caution, Scott not only denied any involvement in the rape but also repeatedly stated that, prior to his arrest, he had never even been to Manchester.

The arresting officers took advice from lawyers representing the Crown Prosecution Service and were told that, as the sole evidence against Scott was the partial DNA profile, they suggested that police conduct further inquiries and find evidence such as bank records or mobile-phone-site analysis to prove that Scott had indeed been in Manchester at the time of the attack.

Despite clearly wishing to see further corroboration, and

despite its own guidelines, which discourage pursuing cases where DNA is the only evidence, permission was granted by the CPS for Scott to be charged with rape. He was denied bail and remanded into custody.

On 12 December the detective constable from Greater Manchester Police who was working on the rape case received details of the location of Scott's phone on the day of the assault. Within a few hours of the attack, the phone had been connected to a cell site in Plymouth, some 340 miles away.

The officer sent an email to the forensics department of the GMP, expressing her doubts and concerns, but the response was delayed, partly due to Christmas coming up and also because it was not possible for the results to be re-checked until an authorization to meet the additional costs had been received.

In the meantime, Scott was jailed for a year in relation to the original street fight. He was no longer being held on remand, but still had the prospect of a lengthy additional sentence for rape to contemplate.

New DNA tests were carried out on 27 February but, rather than going back to the original swabs and taking source material from those, LGC simply re-tested the DNA they had originally extracted. This confirmed the presence of DNA, seemingly from semen belonging to Adam Scott.

Having had the concerns of the GMP detective explained to her, the reporting officer at LGC spoke to colleagues, who suggested that contamination might be a possibility. On this basis, a re-testing of the samples from the original swabs was carried out. On 3 March this produced results that showed no trace of Scott's DNA and, on 7 March, the case against him was discontinued.

An internal investigation at LGC quickly uncovered what

had gone wrong. In March 2011, in order to cope with an ever growing workload, the company had introduced two robotic devices into its Teddington laboratory. The robotic system automated the process of extracting DNA from samples. The samples themselves are loaded on to plastic trays which carry up to 96 in one batch. The trays are supposed to be used once and thrown away, but a junior technician had been reusing them. Some of Scott's DNA remained behind in one of the trays and, when it was used again, it became mixed with the semen sample, leading to the partial profile.

It soon emerged that the misuse of the plastic trays had been spotted on 11 October, just a few days after Scott's DNA had been wrongly identified, and procedures had been changed to make sure the error could not happen again. LGC's mistake had been to believe that no contamination had occurred as a result and to fail to check back to see how long the misuse of the trays had been going on.

Alarm bells should also have sounded on 12 October when a reporting officer at the Risley lab was informed that the control-sample run with the batch containing the GMP rape samples seemed to have been contaminated.

LGC quickly instigated new procedures to ensure that such errors did not occur again and, despite some 26,000 samples having been processed since the introduction of the robots and the time the misuse of the trays was discovered, both the company and the forensic science regulator were satisfied that there were no further incidents of contamination.

A few weeks after the Adam Scott story was made public, LGC was back in the headlines for all the wrong reasons. It emerged that one of its scientists had mistyped a DNA code during the investigation into the death of MI6 code breaker Gareth Williams, whose body was found stuffed into a holdall in the bathroom of his Pimlico flat in 2010. As a result of this, police had spent more than 18 months trying to match a profile that belonged to the scientist himself.

The FSS may be no more, but the same old problems remain.

EPILOGUE

WHEN PRIVATE FORENSIC SERVICES PROVIDERS (FSP) TOOK ON the FSS share of the UK forensic market following the closure of the service in March 2012, a bonanza of profitable work was expected.

Two years before, that share had been worth around £180 million, but by the time the FSS shut its doors for the very last time the value had dropped to just £80 million. Falling crime, increased police force in-sourcing of forensic science examinations and reduced forensic submissions by forces due to funding cuts and changes in technology had all contributed to the dramatic shrink in market value.

When it became an executive agency, the FSS had initiated an approach whereby the police bought 'tests' rather than a service, obliging them to decide what was required before submission to the laboratory in terms of, for example, '3 blood DNA analyses, 1 shoemark examination, 2 glass examinations', and so on. Other providers had followed this lead. This type of charging regime encouraged police to carry out initial screening examinations on exhibits in order to manage their submissions, and hence budgets. Not a good situation to begin with.

Yet perhaps the biggest cause of the collapsing market was the introduction of the National Forensic Framework Agreement (NFFA). This was the brainchild of the short-lived Police Excellence in Procurement Unit, which was set up to coordinate all police procurement, including forensic science, in order to gain the benefits of buying in bulk. The procurement specialists reduced forensic science to a menu of tests to be selected from a huge list, with dire consequences.

Instead of submitting an item of clothing to the laboratory, for example, the police SOCO would cut out or swab a small area which they had tested for blood and send it to the laboratory with instructions to carry out a basic DNA analysis. Any additional intelligence the laboratory would formerly have supplied in the way of the significance of the blood distribution on the clothing was lost. The forensic scientist was being reduced to the role of technician and the 'forensic' element – the independent holistic approach, expert knowledge and in-terpretive skills – undermined.

The NFFA allowed for multiple tenders to be awarded for the same test type so that police forces could play suppliers off against each other even after the contracts were awarded. This resulted in fragmented cases in which items in a single case might be sent to one supplier for DNA analysis, a different supplier for the examination of paint or glass, and yet a third for alcohol analysis. There was no interaction between suppliers and no overview of the value of the forensic evidence as a whole.

Had the case of the Babygro occurred today, only a few small fragments of the garment might have been sent off for analysis and the scientists conducting the procedure would have known nothing about the background circumstances. A presumptive test for spermatozoa would have been carried out, followed by confirmatory examination using a microscope. While it is possible that the presence of vaginal cells might have been noted, it is equally possible that they would not, as the investigation would not have considered that option or requested such information. Certainly the co-location of the semen and vaginal staining (and hence the conclusion that it was due to vaginal drainage) would have been missed.

As a result, the father in the Babygro case could easily have found himself on remand as a potential sex offender, then on trial, with a scientist testifying to the indisputable proof of his guilt as evidenced by the garment. Unless he or his defence team were able to see through his wife's plot, he might well have been found guilty.

Although the FSPs compete with one another to secure work, the biggest potential competitor to forensic science providers in scrambling for the diminishing market is the police themselves. The next three years are crucial. FSPs will have to find new ways of ensuring their worth to forces – especially those forces which are choosing increasingly to in-source.

The problem was predicted far in advance. I, among others, brought this to the House of Commons Science and Technology Select Committee's attention some years ago. The report by the committee, published in 2011, which was highly critical of the decision to close the FSS, warned:

> If the government wants a competitive market in forensic services it must ensure that the market is not distorted by the police customer increasingly becoming the competitor. The government's ambitions for fully privatized forensic science provision are jeopardized by its complacent attitude towards police forensic expenditure.

The stabilization of the shrinking forensics market is now of crucial importance. A shrinking market provides no incentive for further investment or growth from forensic science providers external to the police. The Committee recommends that the government introduces measures to ensure that the police do not further in-source forensic science services that are already available from external providers through the National Forensic Framework Agreement (NFFA) and successor procurement frameworks.

Measures were never introduced.

Apart from the impact on the market, one of the major concerns about the increasing reliance on police in-house forensic services is the quality of the work, especially now that more challenges are being brought against forensic evidence.

All forces in England and Wales are seeking accreditation from the UK Accreditation Service (UKAS). This is to meet European Standard ISO 17025, which endeavours to ensure that in-force forensic examinations and analyses have robust procedures, processes and training in place comparable with those of FSPs. However, as of 2013, only five forces have accreditation to this standard – and then only in some aspects of their forensic work; four are recommended to receive accreditation and twenty-three are still somewhere in the process. But, all the while, they are still delivering unaccredited forensic science casework to the courts. The Catch 22 is that a laboratory has to be delivering the service already, before it can be accredited.

Forces believe that ISO 17025 will prove vital in the courtroom and provide a safety net to in-house police scientists under cross-examination from often intimidating barristers. However, ISO 17025 requires only that forces are delivering forensic examinations to the standard which they set themselves. It does not guarantee that this standard is fit for purpose in a court of law. Furthermore, accreditation demands that suppliers need only inform their customers (themselves in the case of police in-sourcing), not the public, of systemic quality failures.

It is possible that, in the end, the move to a wholly commercial, privatized market will turn out to be a huge success.

It is, however, possible that miscarriages of justice in which people are arrested, tried and convicted on substandard or, indeed, missing evidence will continue to occur. And unless those cases become subjected to additional public scrutiny, no one will be any the wiser.

Like the mythical Hydra of the Bramshill exercises, every question answered seems to throw up two more.

I wonder what might have happened if the examinations and analyses in the Taylor and Nickell cases had been carried out in-force? Would cold case reviews have been sent to commercial FSPs if the in-house laboratory had failed to find anything? With no requirement to inform anyone but the Home Office Forensic Science Regulator, would these failures have come to public attention? Who actually is responsible for ensuring that accurate and reliable scientific evidence is presented to courts?

Being wholly commercial, how stable is the forensic science market in England and Wales? The results of losing a contract to a rival FSP or in-force are dramatic. As a result of the loss of a DNA contract for the Metropolitan Police, LGC announced in March 2013 that it would have to shed between 150 and 170 jobs, including those of scientists and support staff. In its annual report, it stated: 'All of the group's divisions saw growth in turnover, except for forensic science, which saw reduced submissions and casework, principally as a result of pressure on public spending.'

Another main FSP, Key Forensics Ltd, has publicly admitted operating at a loss for the first seven years following its launch in 2005. The move into profit will undoubtedly be assisted by the demise of the FSS.

If even one of the main FSPs chooses to close its forensic division – as well the shareholders might insist – would the market be able to absorb the loss? ACPO has stated that the police forces would use mutual aid to manage the problem. But I doubt it.

In the misconceived and mismanaged drive to a commercial market at all costs, the government has effectively destroyed a jewel in the crown of its criminal justice system. The UK, once world leader in forensic science, is now trailing behind most of Europe. The National DNA Database, once the envy of the world, has now become outmoded, as the rest of Europe introduces the newest DNA technology.

In a shrinking market, where is the incentive for investment in novel forensic science research? Who will fund longer-term development of new techniques and analyses? The government Centre for Applied Science and Technology has a narrow remit for some DNA development, but its resources are far smaller than those the FSS had at its disposal and it relies on expert communities rather than in-house scientists.

England and Wales are now the only countries with a wholly privatized forensic science market – the only countries that literally put a price on justice. Not surprisingly, other countries, including Scotland, Northern Ireland and Eire, are not keen to follow. It may be that England and Wales might, once again, score a first in forensic science: the first country to have to outsource a whole section of its criminal justice system abroad.

And when that happens, I think we will all be losers. Not just the forensic scientists who have lost their jobs and their enjoyment of the uniquely exciting quality of the work itself, but the police, the Crown Prosecution Service, the criminal justice system and forensic science as a whole. But the biggest losers will be the victims of both crime and miscarriages of justice.

And you can't put Humpty Dumpty together again.

ACKNOWLEDGEMENTS

Grateful thanks to Gary Pugh for his encouragement and access to my old case files; his personal assistant Karen Moye for her tireless tracking down of the case files; Jan Eccleston and Sue Woods for their time and helpful suggestions and Bob Green, Roger (pirate) Chapman, Steve Jones and Heidi Zealey for sharing their memories; Sally Gaminara for her support and encouragement.

INDEX

ABO system 25–7, 29, 115, 128
absorption-elution technique 56–7
absorption-inhibition technique 57
acid phosphatese test 41, 158–9
ACPO (Association of Chief Police
    Officers) 170, 174, 279, 320
ACPO National Firearms Database
    180
adenylate kinase (AK) 58, 59, 228
AFIS (Automated Fingerprint
    Identification System) 185,
    194–5
Algeria 248
alien blood 15
Alliott, Mr Justice 151
alpha-amylase test 41
Alsup, Martha Marie 132–6
Andrews, Tommie Lee 118–19
Anguilla, rape and murder case
    132–41
Anthony, Donna 267
anti-human antisera 43
antibodies 25

antigens 25
Applebee, Desmond 161
Arne, Peter 80–3
Asbury, David 186
Ashworth, Dawn 93–7
Association of Chief Police Officers
    see ACPO
Australia
    first court case to involve DNA
        evidence 161
Automated Fingerprint
    Identification System see AFIS
autopsies 44–9, 235
    clinical 45, 235
    collection of fingernails at 235–7
    dissection and examining inside
        of body 46–7
    examining outside of body 46
    forensic 21–2, 45, 46, 235
    removal of brain 47–8

Babygro case 157–61, 178, 316–17
Baker, John 239

ABOUT THE AUTHORS

Mike Silverman began his career as a biologist with New Scotland Yard's Forensic Science Laboratory, specializing in blood pattern analysis, murder and sexual assault investigations. He became one of the first civilian heads of scientific support (CSI and fingerprints) and introduced the first automated fingerprint identification system. His career took him abroad as Head of International Services for the Forensic Science Service and finally to the Home Office as the first Forensic Science Regulator.

Tony Thompson is widely regarded as one of Britain's top true-crime writers. A former crime correspondent of the *Observer*, his major works include *Gangland Britain*, *Bloggs 19*, *Gangs*, *Reefer Men*, *Gang Land* and *Outlaws*. He has twice been nominated for the prestigious Crime Writers' Association Gold Dagger for Non-Fiction, winning the coveted title for his book *The Infiltrators*. He regularly contributes articles to various newspapers and magazines and appears on both television and radio as an expert on crime-related matters.